TYCHO BRAHE

MAPPING THE HEAVENS

TYCHO BRAHE
MAPPING THE HEAVENS

William J. Boerst

MORGAN
REYNOLDS
Publishing, Inc.

620 South Elm Street, Suite 223
Greensboro, North Carolina 27406
http://www.morganreynolds.com

RENAISSANCE SCIENTISTS

Nicholas Copernicus

Tycho Brahe

Johannes Kepler

Galileo Galilei

Isaac Newton

TYCHO BRAHE: MAPPING THE HEAVENS

Copyright © 2003 by William J. Boerst

Library of Congress Cataloging-in-Publication Data

Boerst, William J.
  Tycho Brahe : mapping the heavens / William J. Boerst.
    p. cm. — (Renaissance scientists)
Summary: Presents the life and work of the famous sixteenth-century
Danish astronomer.
Includes bibliographical references and index.
  ISBN 1-883846-97-8 (library binding)
  1. Brahe, Tycho, 1546-1601. 2. Astronomers—Denmark—Biography. [1.
Brahe, Tycho, 1546-1601. 2. Astronomers.] I. Title. II. Series.
  QB36.B8 B64 2003
  520'.92—dc21

                                        2002153640

Printed in the United States of America
First Edition

To Robin,
with love and appreciation

Special thanks to Professor Owen Gingerich for providing information and pictures relevant to the manuscript and to The Royal Library, Copenhagen, for help in locating illustrations of Tycho Brahe and his contemporaries.

Contents

Tycho Brahe

CHAPTER ONE
NOBLE GENIUS

Tycho Brahe decided to become an astronomer in August 1563, when he was sixteen. He had been an amateur stargazer for years, but had not yet made the final decision to devote his life to studying the stars and planets. After this date, he would never again doubt his choice of profession.

The exact date this decision was made is known because Tycho always kept accurate records of his observations. He was a student at Leipzig University in Germany that summer. His family tolerated his love of mathematics and his stargazing hobby, but expected him to pursue a career better suited to a son of the nobility. He studied law during the day, but spent his nights plotting stars and charting planetary orbits. The only observational tool he had was a drafting compass, an angular tool with two arms used to draw circles and

transfer measurements. By aligning his eye with the conjunction of the arms, Tycho would fix the ends of the two arms on two celestial objects and calculate their angular distance. He would then compare his measurements with two sets of planetary tables. One set was based on the ancient system designed by Claudius Ptolemy; the other set was based on the more recent model of the universe proposed by Nicholas Copernicus.

In August of that year, a conjunction, or coming together, of Jupiter and Saturn was predicted to occur. Beginning on August 17, Tycho measured and recorded the distance between the two planets. By August 24, the two planets appeared so close together that they seemed to nearly touch. This conjunction occurred a month earlier than predicted in Ptolemy's planetary tables, and

Ptolemy is depicted in this medieval illustration with a woman who represents the study of astronomy. Ptolemy is holding a small quadrant. The instrument on the bottom left corner is an armillary. (From Margarita Philosophica, 1512.)

Tycho's Denmark

Tycho later claimed it came a day earlier than he had determined it would by using the Copernican tables.

These inaccuracies offended Tycho. He had first been drawn to astronomy because it promised precision on a celestial level. To the young man, it was "something divine that men could know the motions of the stars so accurately that they could long before forestall their places and relative positions." Now he had discovered that there was much work to be done before this divine dream would be a reality. The planets, Sun, Moon, and stars would have to be accurately measured first. Tycho had found his inspiration: his life would be dedicated to mapping the heavens and nothing would be allowed to stand in his way.

Tycho is actually the Latin form of Brahe's given name, Tyge. He was born at Knudstrup Castle in Helsingborg, Denmark, on December 14, 1546. (The region in Denmark where Tycho spent most of his life is now part of Sweden.) His parents, Otte Brahe and Beate Bille (it was common practice for aristocratic women of the sixteenth century to keep their family names) were members of the Danish nobility. Beate had given birth to twins, but only Tycho, named for his paternal grandfather, survived.

Not long after his birth, the first of many controversies that would cloud Tycho's life occurred. Otte's brother Jorgen and sister-in-law Inger had no children. Otte promised his brother that if he and Beate had a second son, Jorgen and Inger could adopt and raise Tycho as their own. One week after Tycho's first birthday, Beate gave birth to a son they named Steen. Jorgen asked that Otte fulfill his promise and send Tycho to live with him, but Otte, who had grown fond of his firstborn, refused. An argument ensued and lasted over a year. Finally, Jorgen and Inger kidnapped two-and-a-half-year-old Tycho and hid him in their castle, Tostrup, in eastern Denmark.

At first the birth parents were furious, but they gradually came to accept the change. After all, Tycho would never have to worry about anything; Jorgen was as wealthy as Otte, and Tycho would inherit all his uncle's property. While living at Tostrup, Tycho traveled frequently to see his birth parents in Knudstrup.

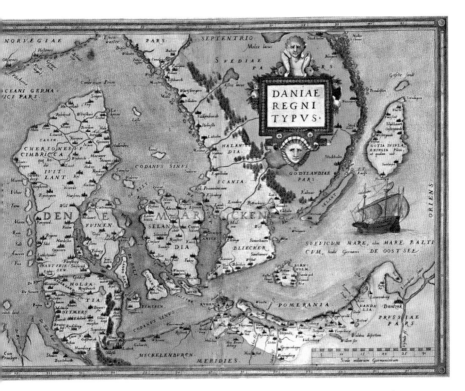

The oldest surviving printed map of Denmark, circa 1570, shows the vast empire during the Renaissance, when it stretched through present-day Denmark, Sweden, and Norway. *(Courtesy of the Royal Library, Copenhagen.)*

Beginning at age seven, Tycho received tutoring in Latin and other subjects, such as horseback riding and fencing. Later he most likely attended a cathedral school near his home until about the age of twelve. At these schools, students and teachers wore robes similar to those of the clergy. The school day began at 7:00 A.M. and lasted until late afternoon, with Wednesday and Saturday afternoons free. The pupils received instruction in Greek, Latin, math, music, and theater.

Students also spent time learning about the new Protestant religion that had been established by Martin Luther, who died only months before Tycho was born. In 1517, when Luther was a young Catholic monk and professor of theology at Wittenberg University in Germany, he listed ninety-five complaints (theses) about abuses in the Catholic Church and posted them on the door of the castle church, which served as the university bulletin board. Within months, Luther was excommunicated (expelled) from the Catholic Church. Although there had been protestors against the Church before, Luther's revolt was the first to receive strong support from princes, kings, and powerful nobles in the German states and elsewhere, mostly in Northern Europe. The result was a rapid spread of the new faith.

The Scandinavian countries were among the first outside of Germany to adopt Luther's ideas on theology and government. Christian III, the ruler of Denmark when Tycho was born, reorganized his country along Protestant lines. He established the same system in Norway, which was then under Danish control.

Tycho lived in an era of rapid social change, conflict, creativity, and destruction. Institutions that had seemed stable for centuries were now in flux. Individuals had to reconsider their relationship to God, to state, and to family. The changes occurring were not as simple as replacing the old ideas with new ones. The intellectual, political, and religious control of the Catholic Church had been broken, and a new world was forming.

In April 1559, twelve-year-old Tycho entered the

Martin Luther started the Protestant Reformation in 1517, when he was a professor at Wittenberg University in Germany. The movement spread rapidly and created an irrevocable schism among European Christians. *(Courtesy of the Library of Congress.)*

University of Copenhagen to follow the usual course of study for children of the nobility, with a concentration on rhetoric (the art of speaking and writing effectively), philosophy, and law. Tycho had only studied Latin grammar at his previous school, but at the university, Latin was the language of instruction and communication between the students, who came from several different countries. An educated man often used the Latin form of his name. Tycho may also have studied Hebrew and Greek. His primary interest, however, was mathematics.

Tycho first became intrigued with astronomy while studying in Copenhagen. He overheard his teachers discussing a solar eclipse that was expected on August 21, 1560, and he found it fascinating that such an event could be predicted. When the date rolled around and the eclipse took place, Tycho was enchanted. He began to turn most of his attention and energy toward astronomy.

He soon discovered, however, that his teachers could not answer most of his questions. He would have to go elsewhere to seriously study astronomy. In the meantime, he began to purchase books, celestial globes, and anything else he could find about the stars and planets.

For almost two thousand years, the writings of the Greek philosopher Aristotle dominated scientific thinking. He advocated the use of employing logic to reach scientific conclusions. He favored this method over conducting experiments or making measurements. As Europe emerged from the Middle Ages, natural philosophers began to use measurement, observation, and cal-

Right: This detail from a Renaissance painting by Raphaël, *The School of Athens,* shows the Greek philosophers Plato *(left)* conversing with Aristotle *(right)*.

Ptolemy compiled his work on mathematical astronomy in a book called *Almagest*. The manuscript was translated by various scholars throughout the Middle Ages. This page is taken from a fourteenth-century manuscript. *(Courtesy of the British Library.)*

culation to test ideas and theories. Modern science was being born; it was an exciting time for a brilliant boy with an aptitude for mathematics.

Aristotle, whose ideas were still dominant in most universities, had laid out some basic astronomical principles in a work, *On the Heavens*, written around 350 B.C. He said Earth was at the center of the universe and the planets, the Sun, and the Moon revolved around it. His later followers determined the planets traveled on crystalline spheres in uniform circular orbits. The entire universe was contained within an outer wall of unchanging stars. Aristotle believed uniform circular orbits were expressions of perfection. Although Aristotle wrote before the Christian era, his ideas were adopted by the new religion and became embedded in Christian theology.

There were problems with the geocentric (Earth-centered) model from the beginning; most concerned the motions of the planets. The word planet comes from the Greek word for "wanderer," and it was the tendency of planets to not maintain a uni-directional speed among the stars that created the most problems for astronomers. Mars, for example, occasionally appeared to stop and move backward, then hesitate again before continuing on its proper way. This phenomenon is called retrograde motion.

In the second century, a follower of Aristotle, Claudius Ptolemy, a Greek who lived in the Egyptian city of Alexandria, designed a system that used mathematical formulas to predict planetary movement. In his model, each planet orbits Earth on a large circle called a deferent. In turn, the planet sits on an epicycle that moves on the deferent.

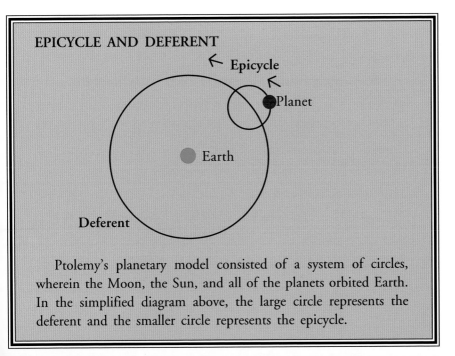

EPICYCLE AND DEFERENT

Epicycle

Planet

Earth

Deferent

Ptolemy's planetary model consisted of a system of circles, wherein the Moon, the Sun, and all of the planets orbited Earth. In the simplified diagram above, the large circle represents the deferent and the smaller circle represents the epicycle.

Ptolemy devised a system that maintained Aristotelian laws and provided a reasonably accurate method to predict the planets' locations and movements. Over the centuries, further refinements were made to Ptolemy's planetary system. Although highly complicated, it held sway over astronomy for fourteen hundred years. It upheld the Aristotelian dictates enough to be adapted into the Christian world view. It was not until a Polish bureaucrat in the Catholic Church decided, as he neared his death in 1543, to publish his life's work, *On the Revolution of the Celestial Spheres*, that a concerted effort to develop a new system was begun.

Nicholas Copernicus is today often viewed as a revolutionary who sparked the "scientific revolution." Actually, this quiet and devout churchman was motivated to offer an opposing planetary system because he was offended by Ptolemy's manipulation of Aristotle's principle of uniform motion. He was determined to find a way to prove planets orbited in circles without resorting to the use of equants.

The problem that dogged Copernicus and Ptolemy before him is planets do not travel in circular orbits. This is one of the discoveries Johannes Kepler would make almost a century later. To do away with equants, and maintain circular motion, Copernicus had to alter something else. He decided to move the Sun. By placing the Sun in the center, and having Earth orbit the Sun while also spinning on its axis, he was able to locate the planets in order. This also meant it should be possible to measure the distance between the planets. Copernicus attributed

Copernicus placed the Sun in the center of the planetary system.
(From Cellarius, Atlas Coelestis, *1660. Courtesy of The British Library.)*

the retrograde motion of Mars to its orbital location relative to Earth's. Mars, for example, is further away from the Sun than Earth. Combining his belief in uniform motion, and the fact Mars travels a greater distance than Earth, which has an "inside lane," Copernicus said retrograde motion occurs when Earth "laps" Mars, a natural consequence of its shorter, inner orbit.

Placing the Sun at the center of the planetary system, and placing Earth on an orbit relative to the other planets, answered several questions and began the movement toward an accurate understanding of the solar system. Unfortunately, Copernicus's loyalty to the Ar-

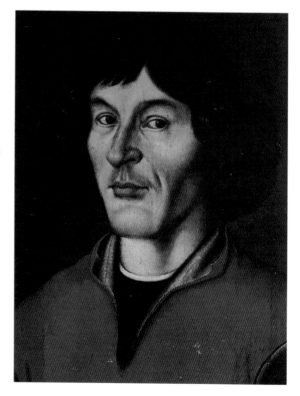

Nicholas Copernicus proposed a heliocentric model of the universe, or one in which the Sun is at the center, in his book, *On the Revolution of the Celestial Spheres.* (Courtesy of the District Museum, Torun.)

istotelian features of uniform circular orbits forced him to resort, as had Ptolemy, to epicycles and deferents. It would be the next century before the heliocentric system would be widely accepted as more than a hypothesis. The process would begin when Johannes Kepler used Tycho's data to determine that orbits are elliptical—not circular—and that a planet's speed varies relative to its distance from the Sun.

Despite its limitations, Copernicus put forth the first serious attempt to remake the cosmos in almost fourteen hundred years. His ideas slowly spread among universities and attracted the attention of intellectuals. When Tycho began his education, Copernicus's ideas were hotly debated and the young Dane decided, at first, to withhold judgement. While he did find the Copernican system a better predictor of planetary motion, several aspects of it troubled him. The elements of the Copernican system that most worried Tycho were the ones that most clearly contradicted Aristotle's writings on physics and motion.

Aristotle had said there were only four elements—earth, fire, water, and air—and earth was the heaviest. It seemed improbable to Tycho that the heavy Earth could hurdle through space, while the lighter Sun remained stationary. Then there was the problem presented by Earth spinning on its axis. If Earth were spinning rapidly enough to make one rotation every twenty-four hours, would an object dropped from a high point not fall to earth at an angle?

PARALLAX

The phenomena of parallax is defined as the apparent shift in position of an object when viewed from different vantage points.

Parallax can be observed on a small scale by looking at an object (a clock, for instance) from across a room. Closing your right eye, hold a pencil at arm's length so that it blocks your view of the clock. Now, close your left eye and open your right eye. The apparent shift of the pencil represents parallax.

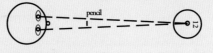

Finally, if the universe were small and compact, as Aristotle had said, it should be possible to detect changes in the size and location of the stars as Earth orbited the Sun. This changing of size and location of an object, relative to its distance from the observer, is a phenomena called parallax. The inability of astronomers, including Tycho, to detect parallax was one of the principal reasons they were skeptical of the Copernican system. Today we know that the universe is countless times larger than Aristotle or Copernicus thought. The distance from Earth to the stars is so great, and Earth's orbit so comparably insignificant, that parallax was not detectable until more powerful telescopes were built in the nineteenth century. In the meantime, this inability to detect stellar parallax provided a powerful argument against heliocentricism.

Until further advances were made, the Copernican

system was restrained by Aristotelian physics. Fortunately for the future of astronomy, Tycho, a man of action, did not let these questions interfere with his determination to measure the skies. He was always most interested in mapping planets and measuring stellar and solar positions. It was this dedication to observational astronomy that led to the collection of a vast resource of data that would later be used to create the science of physical astronomy.

CHAPTER TWO
STUDENT DAYS

During his three years in Copenhagen, Tycho spent more time studying math and astronomy than the prescribed courses of rhetoric and philosophy. He pored over his astronomical tables and Copernicus's *Revolution of the Celestial Spheres,* while befriending other scientists and intellectuals.

Tycho's uncle and adoptive father, however, had different plans for his son. In 1562, Jorgen sent Tycho to Leipzig University in eastern Germany to study law. Andrew Sorenson Vedel, four years older than Tycho, went along as his companion and tutor. (Vedel would later become Denmark's first great historian and collector of Nordic sagas.) Tycho and Vedel left Copenhagen on February 14 and journeyed by ship and horseback to Leipzig. After arriving on March 24, they boarded at the home of a professor. Tycho spent much of his time in

Leipzig beseeching professors of mathematics to teach him.

Vedel had been assigned the task of keeping Tycho away from astronomy, and he tried valiantly, yet vainly, to do so. There was no stopping his young charge, who was becoming increasingly obsessed with mathematics and medicine, as well as stargazing. Tycho's family hoped he would see the error of his ways and turn his attention to more useful pursuits. In letter after letter, his family begged Tycho to not forget his family responsibilities.

Tycho faced other obstacles to becoming an astronomer. Before the invention of the telescope around 1610, making accurate astronomical observations was difficult and frustrating work. The available instruments were crude and expensive, and their use required good eyesight and a steady hand. Later, Tycho would devote a great deal of his fortune to designing and building larger and more accurate instruments, but as a seventeen-year-old student his first tool was a pair of ordinary drafting compasses. To make measurements between two objects, such as two stars or a planet and a star, Tycho would hold the compasses close to his eye and direct the legs at the two points. Tycho and the two points of the compass formed a triangle. This allowed him to use trigonometry, the branch of mathematics that deals with the relationships of the sides and angles of triangles, to determine the angles between the objects and plot the relative positions of the stars and planets.

Tycho's observations of the conjunction of Saturn and Jupiter in 1563 spurred him to accumulate more instruments. The following year, he acquired a cross staff, a tool used mostly in navigation. The cross staff consists of two graduated rods that form a right angle and is used to measure the angle of elevation of an object in the sky. Tycho's first observation made with a cross staff occurred on May 1, 1564, while Vedel was sleeping. Tycho also owned a celestial globe that he studied to memorize the names of the constellations. Even better, he could easily hide the fist-sized globe until Vedel gave up trying to control Tycho's studies. Vedel grew to value his young charge's intelligence and vigor. The two remained lifelong friends.

In mid-May 1565, after three years at the university in Leipzig, Tycho and Vedel returned home to Denmark. Denmark and Sweden were at war, and Tycho's adoptive father, Jorgen, was vice admiral of the Danish fleet. In June, an accident caused King Frederick II and Jorgen Brahe to lose their balance while crossing a bridge near the royal castle at Copenhagen. (Some sources reported the two had been drinking.) The king fell in the water first, and Jorgen was able to rescue him. Soon after, Jorgen developed pneumonia and died. After Jorgen's death, Tycho's birth parents took over their nineteen-year-old son's affairs.

Tycho stayed in Denmark almost a year. Then, early in 1566, he and Vedel left for more schooling in Wittenberg, Germany—the birthplace of the Protestant

Andrew Sorenson Vedel was appointed as Tycho's tutor. The two men would
come to respect one another as equals and remain lifelong friends. *(Courtesy Det
Nationalhistoriske Museum på Frederiksborg, Hillerød.)*

CROSS STAFF

A cross staff is a simple tool dating back to ancient times. Its cross shape is used to determine the angle between two objects. Before the invention of the telescope, the cross staff was a vital tool in astronomy, and Tycho Brahe was one of the last great astronomers to use it.

The cross staff consists of a long arm (staff) and a perpendicular shorter arm (cross piece) that can be slid up and down the staff. The viewer places one end of the long arm against the eye, with the staff pointing outward. The viewer moves the cross piece until each end of this shorter bar lines up with the two objects in question, such as two celestial objects, or the horizon and a heavenly body.

The ratio of the length of the cross piece to the length measured on the staff provides the data necessary to determine the angle. If the longer arm were marked with a graduated scale, the viewer could determine the angle between the two objects by noting the point where the staff and the cross piece intersect. If the staff were marked with a linear scale, the ratios would be calculated using Trigonometry.

Reformation and one of the leading intellectual centers in Central Europe. Tycho wrote a friend about his decision to leave Denmark: "Neither my country nor my friends keep me back. One who has courage finds a home in every place and lives a happy life everywhere. Friends too, one can find in all countries. There will always be time enough to return to the cold North to follow the general example, and like the rest, to play in pride and luxury for the rest of one's years with wine, dogs, and horses."

There may have been a less romantic reason for Tycho to leave his homeland. Although he found support with his uncle Steen Bille, and Vedel had secretly been won

This woodcut from 1533 shows how the cross staff could be used to measure the distances between two points, both on Earth and in the sky. (*From J. Werner and P. Apianus:* 'Introducio Geograhica' in Doctissimas in Veneri Annotationes, *1533.*)

over, his other relatives still did not approve of his stargazing.

Tycho and Vedel arrived at Wittenberg on April 15, 1566, where they both attended classes at the University of Wittenberg, but not for long. After five months, plague broke out, and the two fled to the university in Rostock, Germany.

Unlike Wittenberg, no noteworthy astronomers lived in Rostock, but Tycho did witness a lunar eclipse there on October 28, 1566. That winter, he and Vedel decided to stay in Rostock for Christmas. He had made friends with a theology professor named Lucas Bachmeister and wanted to spend more time discussing their mutual interest in science and mathematics.

During the Christmas season of 1566, Tycho attended a party at the professor's house. As the party progressed, and ample quantities of ale were consumed, Tycho began quarrelling with another Danish student named Manderup Parsberg over who was the most gifted mathematician. The drunk students almost came to blows before their friends intervened. The following week, on December 29, they met at another party, again drank too much, and renewed the quarrel. This time violence could not be avoided. The night's events were told to Tycho's first biographer by the descendant of a witness to the events:

> [Brahe] unexpectedly got into an argument with one of the table companions [Parsberg], and soon they were so wrought up, speaking in the Danish language, that they demanded swordplay of each other, stood up forthwith, and went out. My late grandmother, who knew the Danish language and was eating in that same room, admonished the other table companions to follow them straightaway and to try to hinder any misfortune, which they indeed did do. But when they came out into the churchyard, the others were in full brawl, and Tycho had received a stroke that had hacked away his nose.

This picture shows clearly the scar left after Tycho lost his nose. *(Courtesy of The Royal Library, Copenhagen.)*

Losing most of his nose became the most talked about event in Tycho's life, but he refused to let the disfigurement keep him out of society. A friend explained: "As Tycho was not used to going around without a nose, and did not like to, he went to the expense of purchasing a new one. He was not satisfied, as some others might have been, to put on a wax one, but, being a nobleman of wealth, ordered a nose made of gold and silver so soberly painted and adjusted that it seemed of a natural appearance." For the rest of his life, Tycho could often be seen rubbing an ointment or glue onto his nose. Surviving portraits show an unattractive scar across his forehead and a line running across the bridge of his nose.

This fifteenth-century depiction shows the influence of the zodiac *(right)* on the human body *(left)*. It was believed the astrological signs influenced health as well as behavior. *(Courtesy of the British Library.)*

At this time, most people believed that the positions of the stars and planets could foretell natural phenomena, such as earthquakes or drought, and such human affairs as a change of political power or a person's character and fate. Mathematicians and astronomers received money from wealthy patrons who desired to have their newborns' birth charts made or their own fortunes told. Therefore, understanding astrology was considered to be a practical skill, and it provided Tycho with a justification for his stargazing.

Although Tycho professed to believe in the power of

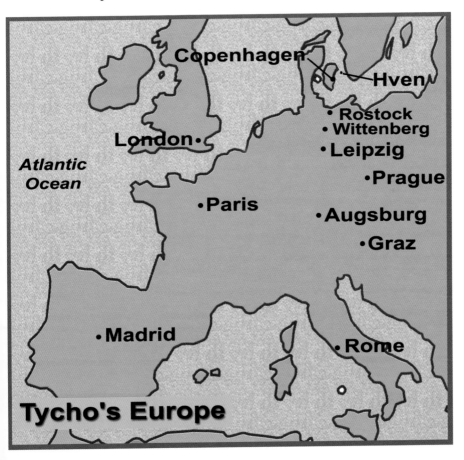

heavenly bodies to influence earthly events, he warned against the misuse of astrology and argued that God could change events at any time. During the previous October's lunar eclipse, he tried his hand at public astrology and predicted the death of the Turkish sultan. Unbeknown to Tycho, the sultan had died almost six weeks earlier. Apparently, he had been too busy with his studies and stargazing to keep up-to-date with current events. He would continue the practice of astrology throughout his life at the request of princes and kings.

Despite his self-confidence and determination, Tycho was susceptible to homesickness. After visiting Den-

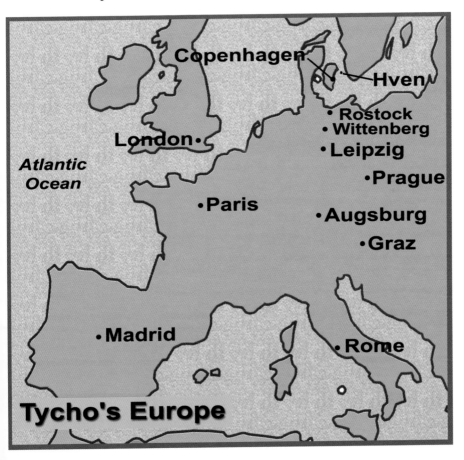

Copenhagen

Hven

Rostock
Wittenberg
London
Leipzig

Atlantic
Ocean

Prague

Paris

Augsburg

Graz

Madrid

Rome

Tycho's Europe

SEXTANT

Although the sextant was known to Islamic astronomers, Tycho Brahe made it his signature astronomical instrument. He made several modifications over the years. The sextant was similar in purpose and design to the quadrant, both are designed to determine the angular separation of two celestial objects, as well as measuring the altitude of a celestial body. The sextant is a sixth of a circle, versus a quarter circle of the quadrant.

Its smaller size, marking sixty degrees, could be more portable than a quadrant. As Brahe wrote, this could be an advantage because "an astronomer, more than the student of other branches of knowledge, has to be a citizen of the world, and consider every place to which circumstances or necessity might lead him as his native country."

mark in late 1567, he lamented to a friend, "I was better received in my native land by family and friends than I deserved; the only thing lacking was that everybody be pleased with my studies." He would have liked to remain in Denmark, but there was little opportunity there to be an astronomer. Instead, he moved on to the University of Basel, Switzerland, and the following year to the University of Augsburg, Germany.

It was common for a young nobleman to move from one university to another. Because their futures were usually assured in the royal court or in the military, noblemen had no need for degrees, although an adequate education was important. Such mundane tasks as completing degrees, writing books, and working in science, were left to the middle class.

In Augsburg, Tycho became friends with astronomer Paul Hainzel. Paul and another friend were convinced they needed a large quadrant to make more accurate astronomical measurements. The quadrant is a simple instrument, a graduated quarter of a circle that is used to measure the altitude of objects above the horizon. A larger quadrant had space for more precise degree gradations, which allowed for the marking of not only the ninety degrees of a quarter circle but also the minutes between each degree mark.

Hainzel agreed to pay for a nineteen-foot quadrant made of oak. Augsburg craftsmen, supervised by Tycho, completed it in a month. It took twenty men to erect it on a hill in Hainzel's garden, where it remained five years and was used to make observations until destroyed by lightening in a storm in 1574. Tycho also began to use a sextant, an instrument similar to the quadrant and used to measure angular distance. He oversaw the building of a five-foot celestial globe as well, but this was not finished by the time he left Augsburg.

Tycho also became friends with Petrus Ramus, a professor of philosophy and rhetoric. Ramus was a French Huguenot, or Protestant, who had fled his homeland to avoid persecution. Ramus was a progressive thinker who rejected Aristotle. He believed that the researcher should begin without any preconceived ideas, or even a hypothesis to be tested. Tycho disagreed and said astronomy without a hypothesis was impossible, but he did agree there was a need for more objective research and data.

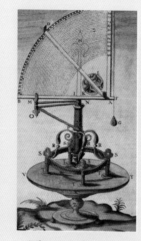

QUADRANT

The quadrant consists of a quarter circle, a movable arm, and a plumb line. It was used to measure the altitudes of stars and planets. The curve of the quadrant is graduated with ninety degrees and positioned so its plane passes through the celestial object in question. The plumb line guarantees that one side of the quadrant is horizontal, and the other pointing toward the zenith, or the point in space that is directly above the viewer. The movable arm is used to site the object.

The quadrant can be traced back as far as Ptolemy, and was used extensively by Islamic astronomers. Quadrants were used even after the discovery of the telescope, because early telescopes could not measure exactly where a star was. There are two types of quadrants, mural quadrants, which were fixed to a wall, and altazimuth quadrants, which could be rotated. Brahe employed both.

According to Brahe, "The use of this quadrant is for finding the altitudes of the stars and of the sun and moon in such cases when complete accuracy is not necessary."

Early in his career, Brahe designed a nineteen-foot wooden quadrant for Paul Hainzel of Augsburg, Germany, graduated into degrees and minutes. As there are sixty minutes in a degree, the quadrant had 5,400 gradations. After only five years, this quadrant was struck by lightning during a storm and burned down.

Although he did not always agree with Ramus, Tycho admired his fresh approach.

Tycho received word that his father was ill and returned to Denmark in 1570. He left Paul Hainzel in charge of the construction of the large celestial globe. At home he discovered that the life of a nobleman—the frequent travel, heavy drinking, and constant pressure—had taken its toll on his fifty-three-year-old father. Although his father survived the winter, he died in early spring at Helsingborg Castle.

Otte Brahe left several properties to be divided among his wife and sons. (Daughters were not entitled to an inheritance.) Together, Beate and her sons, Tycho and Steen, became co-administrators of Knudstrup. The estate consisted of 322 farms, twenty-nine cottages, and seven mills. Tycho and Steen's share came to two hundred farms, twenty cottages, and five and one half mills. Otte had left behind other properties to be divided among his heirs, including five hundred farms, sixty cottages, and fourteen mills throughout Norway and Denmark. There were also four residences and more than forty rental properties in Copenhagen and other cities. Income from timbering and the raising of swine came from forest areas in Jutland, Denmark. In total, Tycho now commanded an immense fortune that would have been enough to satisfy most men, but he dreamed of building the world's most magnificent astronomical observatory. He was even willing to leave his native land in order to make his dream become a reality.

CHAPTER THREE
STARGAZER RECEIVES AN OFFER

Although Frederick II, the king of Denmark from 1559 to 1588, had dyslexia, a disorder that makes reading difficult, he loved the world of ideas. He established scholarships to educate his poorer subjects and looked for ways to attract scholars and professors to Denmark. As Tycho became well known in scholarly circles throughout Europe, Frederick, whose life had been saved by Tycho's adoptive father, wanted to keep the young scholar in Denmark. He promised Tycho the canonry, a supervisory position, at Roskilde Cathedral, which provided a salary and involved little work. There was one hitch in the deal, though. Tycho would have to wait for the present canon's death before he could assume the office. As he considered the offer, Tycho decided to stay in Denmark.

Tycho, the oldest male of one of Denmark's most

prominent families, began to spend more time at royal court. He became better acquainted with King Frederick II, who was twelve years older. The two men discovered that they had had a common experience—they both had fallen in love with women from the lower class.

Tycho met Kirsten Jorgensdatter, whose father, Jorgen Hansen, was a Lutheran minister, in the weeks while his father was dying. Unfortunately, ministers were not nobles, and relations between the two classes were tightly restricted by laws governing property and inheritance. For Tycho to seriously pursue a marriage with a commoner was yet another insult to his family. Characteristically, he did not let this deter him. Eventually, he and Kirsten would live together in a common-law marriage. As the law explained, "The woman who for three winters lived openly as wife in a house, eating and drinking and sleeping with the man of the house and possessing the keys to the household" was that man's wife.

King Frederick was sympathetic to Tycho's plight because he had fallen in love with a woman who, although a member of the nobility, was not of royal blood. Any children born of their union would not be able to inherit the throne. Frederick resisted the pressure to marry a member of the royalty until he was thirty-eight. Then, in July 1572, he married his fourteen-year-old cousin, Princess Sophia of Mecklenburg.

The summer wedding brought people from all over Europe to the ceremony in Copenhagen. Sophia was crowned queen the day after the wedding. During her

This bust of Frederick II
was created in 1577.
*(Courtesy Det
Nationalhistoriske Museum på
Frederiksborg, Hillerød.)*

coronation, Kirsten's uncle Peder Oxe carried the crown,
and Tycho's mother was one of the new queen's ladies-
in-waiting. In celebration of the event, there were many
banquets, parades, tournaments, plays, and masquer-
ades. Shortly after their wedding, the royal couple produced an heir.

Tycho and Kirsten moved in with his uncle Steen Bille

at Herrevad Abbey, twenty miles east of Helsingborg. Tycho had a great deal in common with Steen, who had built the first paper mill in Denmark and spent a great deal of time in his alchemist laboratory. The study of alchemy was an art that had its roots in ancient Egypt. Reintroduced to Europe through the Moors in Spain during the Middle Ages, the search for an elixir that could transmute ordinary metals into gold was not limited to crackpots or aspiring wizards. The philosophical assumptions underpinning alchemy—the dynamic nature of the physical world, the changeability of elements through the application of chemicals, and the necessity of sensitive and precise experiments and careful observation during experiments—were incorporated into modern science.

Steen had ample space to house Tycho while he undertook his astronomical observations. On the evening of November 11, 1572, as Tycho walked back to the house from his lab, he looked at the sky and saw a light shining near the constellation Cassiopeia, where none had existed before. Tycho could not believe his eyes and called out to nearby servants to verify what he saw.

Other astronomers saw the phenomena—the light continued to shine for a month—and some claimed that it was so bright it could be seen during the day. But what was it? According to Aristotle, this could not be a new star, suddenly appearing in the fixed outer sphere of heaven, as change was only possible between Earth and the Moon in the sublunary sphere.

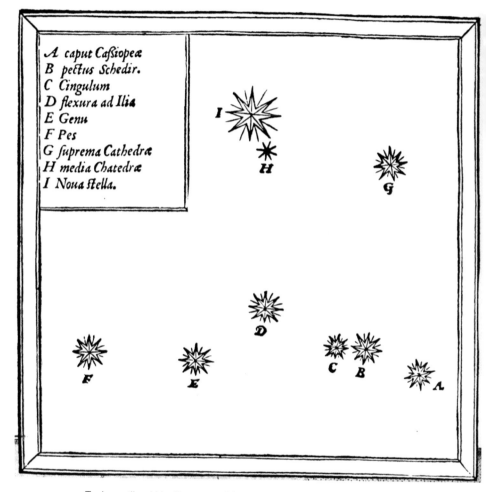

A caput Caßiopeæ
B pettus Schedir.
C Cingulum
D flexura ad Ilia
E Genu
F Pes
G fuprema Cathedræ
H media Chatedræ
I Noua ftella.

Tycho outlined his discovery of the new star in *De stella nova*, including this map that shows the position of the star in the night sky. The new star is labeled I, at the top of the drawing. *(From Tycho Brahe:* De stella nova, *1573.)*

There was only one way for Tycho to determine if what he saw was a star. If the light was fixed, it was a star; if it moved, it was not. Fortunately, Tycho had recently finished building a huge sextant with sides five and a half feet long. For eight months he took frequent sightings of the mysterious light, often several each night, precisely recording its location. Satisfied with the data he had collected, Tycho concluded it was indeed a star in the sublunary sphere. He was convinced a new star had been born. Today we know he actually witnessed a supernova, an exploding star, which occurs at the end of a star's lifetime when its nuclear fuel is exhausted and it is no longer supported by the release of nuclear energy.

Tycho was reluctant to publish his findings. Finally, Peder Oxe, Kirsten's uncle and an amateur astronomer, prevailed upon Tycho to put his ideas into print. In 1573, Tycho published presentation copies, a sort of pre-publication edition that did not contain the elaborate illustrations of a finished work, entitled *De stella nova* (*The New Star*). The work explained his observations that confirmed a new star had indeed appeared. He also used the book to emphasize the need for repeated, systematic direct observation in astronomy. Tycho ended the book with a poem explaining that his interest in astronomy was his way of trying to understand nature. *The New Star* established Tycho's reputation as a scientist. Johannes Kepler later said this new star had also heralded the arrival of a great astronomer.

As part of his continuing effort to keep Tycho in Denmark, King Frederick asked him to lecture at the University of Copenhagen. Tycho agreed to the king's request, although it was frowned upon for noblemen to undertake such a mundane position. At the university, he lectured on astronomy and astrology. According to Tycho, studying astronomy was a practical necessity because it is used to divide time into years, months, and days. It was "of such value that a knowledge of it must, for good reason, be desired by every noble and high-minded person. It fills one with tremendous joy and sharpens one's understanding."

With the added job of teaching, his many responsibilities often kept him away from his own work. Tycho, who dreaded living a life of "horses, dogs and luxury," avoided as best he could engaging in social activities in the bustling city of Copenhagen. He complained about the unfairness of being part of the nobility, stuck in "feats of arms, concourse with kings and princes, and the pursuit of wine, women, and song." He preferred the world of books and learning at the university, although teaching classes took up too much of his time. After his father's estate was cataloged and appraised, Tycho discovered that he would be receiving 650 dalers per year, compared with a professor's salary of three hundred dalers per year. He decided to give up teaching and focus on his scientific research and other work.

Although Tycho had been stigmatized for becoming an astronomer and for entering into a common-law marriage with Kirsten, he had become a highly respected figure. The publication of *The New Star* had even brought him some fame. Despite King Frederick's attempts, he was growing discontented with life in Denmark. He wanted a more intellectually active life. His dream was to run an observatory, collecting and organizing data. He began to think this goal could best be achieved in one of Europe's intellectual centers, such as Wittenberg or Basel. He began preparing to leave with his family for the continent, this time for good. His plans were interrupted, however, when he suffered a bad attack of *ague* (repeated cycles of fever and chills) in the summer of 1573.

Tycho's first child, a daughter named for her mother, was born in 1573, and a second daughter, Magdalene, was born in 1574. His female children were given the last name Tygesdatter, meaning "Tyge's daughter."

After the births of his daughters, Tycho finally embarked on his long awaited trip. Leaving his family at home, he began a tour of Europe in order to find the best place to move. He first visited Landgrave Wilhelm IV of Hesse-Cassel. (Landgrave was a German title of nobility corresponding to an earl in England or a count in France.) Wilhelm had an interest in astronomy and understood the value of regularly recorded observations. Fourteen years earlier, he had built his own observation tower. During Tycho's visit, the two talked by day and observed the sky by night. At the end of the week, the death of Wilhelm's daughter called him away, but the

two would correspond and compare notes for the rest of their lives.

Tycho's next stop was Frankfurt, Germany, where he purchased books at a large book fair. He then went to Basel, in present-day Switzerland. Basel was an intellectual capital of Europe and was centrally located, which would make it easier to travel to other cities. It seemed to be the perfect place to build his observatory. He returned to Denmark with the intention of moving to Basel, where he could focus all of his attention on astronomy.

CHAPTER FOUR
HIS LORDSHIP OF URANIBORG

When Tycho returned to Denmark, his plans to move to Basel secure, King Frederick surprised him with another attempt to keep the astronomer in Denmark. Frederick offered Tycho a choice of castles at which to set up an observatory and continue his work. At first, Tycho refused the offer, as he was determined to leave for Basel. Then, before dawn one morning in 1575, a courtier appeared at Tycho's door with a letter from the king requesting that he come immediately. The two men met privately and what Frederick said startled Tycho. He later recorded the king's words:

> One of my courtiers reported that your uncle, Steen Bille, told him in secret that you are planning to return to Germany. Therefore, I wanted to summon you to me, so I could hear from you yourself what you think of my proposal and why you have not accepted it. I

suspect that you do not want to accept a great castle as a sign of royal favor because the studies you enjoy so much would be disturbed by the external affairs and concerns required to maintain and command a castle, and because you do not want to neglect your learned investigations.

Then Frederick told him he had another proposal for Tycho to consider:

I saw the little island of Hven, lying in the Sound . . . It occurred to me that it would be very well suited to your investigations of astronomy, as well as chemistry, because it is high and has an isolated location. Of course, there is no suitable residence, and the necessary incomes are lacking as well, but I can provide those things. So if you want to settle down on the island, I would be glad to give it to you as a fief. There you can live peacefully and carry out the studies that interest you, without anyone disturbing you . . . What good would it do for you to return to Germany and be a stranger there, when you can accomplish every bit as much in your native land?"

The king offered Tycho solitude on his own island— if he agreed to stay in Denmark. Tycho thanked him and promised to think it over. Back at Knudstrup, he wrote to his friend Professor Johannes Pratensis at the University of Copenhagen, asking his advice. Pratensis wrote back, "Apollo desires it, Urania recommends it, Mercury commands it." Tycho went to Herrevad Abbey and dis-

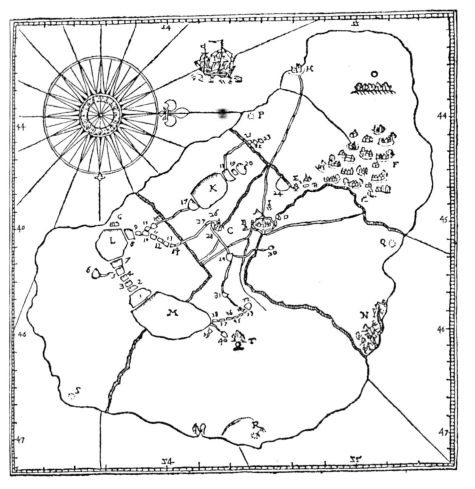

The island of Hven is located in Øresund, the sound that lies between present-day Denmark and Sweden. *(Willem Janszoon Blaeu:* Epistolarum astronomicarum, *1596.)*

cussed it with his uncle. He also talked to Kirsten, his mother, and his brother Steen.

On February 18, 1576, Tycho pledged to work at Hven for a salary of five hundred dalers a year. Four days later he was on his island, "with its white cliffs rising steeply out of the sea." Hven was fourteen miles north of Copenhagen, nine miles across the sound from the king's castle, Elsinore.

The island was about two thousand acres. At its longest, it measured three miles from end to end, and it was about one and a half miles wide. The perimeter measured seven miles. There were approximately forty farms on the island, and the southern two-thirds was maintained for grazing land.

The inhabitants of Hven, about two hundred people, lived in a village called Tuna near the north end of the island. Men wore beards and long trousers with belted smocks, ruff collars, and high hats with small brims. Women wore long-sleeved blouses, long wool dresses, and full-length aprons. Married women donned linen caps over their hair, while maidens braided their hair. Children dressed much like the adults.

Tycho, on the other hand, sported a long moustache over his blond beard and usually wore knee breeches, a buttoned vest, linen shirt, lace ruff (a high, ruffled collar) and cuffs, a high-collared cape, and gloves. On his head, he perched a plumed beret, and a sword swung along his thigh.

The disparity in dress reflected the social and political differences between Tycho and the islanders. Tycho considered himself to be a royal vassal, subject only to the king, who had given him control of the island. He had the authority to collect rents and demand free labor. The only cost to him, mandated by law, was the free meal he had to provide to the peasants while they worked, and a slight tax he paid the king. Tycho also assumed control of the island's church and its clergy, which entitled him to two-thirds of all tithes, typically one-tenth of the crops produced on the island.

The Lord of Hven wasted little time before getting down to work. The first recorded observation in Hven, the conjunction of the Moon and Mars, took place on February 22, 1576. Three months later, the king's gift became official, and the island of Hven, including authority over its peasants and servants, belonged to Tycho for life. Frederick also gave him four hundred dalers for the construction of a new home. The formal proclamation read:

> We, Frederick the Second, &c., make known to all men, that we of our special favor and grace have conferred and granted in fee, and now by this our open letter confer and grant in fee, to our beloved Tyge Brahe, Otte's son, of Knudstrup, our man and servant, our land of Hven, with all our and the crown's tenants and servants who thereon live, with all rent and duty which comes from that, and is given to us and to the crown, to have, enjoy, use, and hold, quit and free, without any rent, all the days of his life, and as long as

he shall keep the tenants who live there under law and right, and injure none of them against the law or by any new impost or other unusual tax, and in all ways be faithful to us and the kingdom, and attend to our welfare in every way and guard against and prevent danger and injury to the kingdom. Actum Frederiksborg the 23rd day of May, anno 1576. Frederick.

The arrangement between King Frederick and Tycho represented the long-established patronage system that existed between royalty and nobility. Grants in fee, gifts from the king, could only be given to members of the nobility, and much of the king's control of the nobility depended on his power to make such grants. Some of the king's gifts carried monetary value, but others did not.

Both patron and client benefited from the arrangement. Tycho was able to realize his dream of having a central research facility where he could make astronomical observations. The king knew other nations would look with favor upon his project; this would help him attract other scientists and intellectuals to Denmark.

Although patronage was not uncommon, an arrangement on this level was. Frederick's gift to Tycho was probably the most generous state contribution to scientific research of the era. Other kings and nobles paid for mathematicians to be on staff or provided assistance to help a researcher continue working, but the gift of an entire island, and the funds to build a large home and an observatory, was uniquely Tycho's.

Tycho Brahe took possession of the island of Hven in 1576. *(Tycho Brahe:* Astronomiæ instauratæ mecanica, *Wandsbek 1598. Courtesy of the Royal Library, Copenhagen.)*

From the beginning, tension existed between the new lord and the farm families on Hven. The peasants were used to handling their own affairs and for decades had considered themselves to be free landholders, subject only to the king. They felt no loyalty to this nobleman, who intended to run the island like a huge estate. Many complained about overwork; the more desperate ones sailed away rather than accept the new conditions. When Tycho complained about the fleeing tenants, the king forbade the residents to leave without permission.

Tycho found a site in the middle of Hven to build his

new home, Uraniborg (literally "castle of Urania," after the goddess of astronomy). On August 8, 1576, a group assembled at sunrise, when Jupiter and the Moon were setting, to lay the foundation stone. There were toasts and the drinking of wine as the stone was placed at the southeast corner of the foundation.

The master mason, Hans van Steenwinkel, had originally come from Antwerp, Belgium. He would also work on King Frederick's country estate during the winters of 1577 to 1581. When he could be spared from his work for the king, Tycho brought him to Hven, taught him the basics of astronomy and geometry, and put him to work building the house.

Although the house took four years to build (it was not finished until 1580), Tycho and his family took up residence before completion. The house, outbuildings, and grounds were magnificently designed. Workmen excavated sixty holes for the fish ponds Tycho ordered.

MURAL QUADRANT

In 1582, Tycho had a large quadrant built onto the wall of his study with a mural depicting Tycho and his assistants painted inside its arc. This was one of the largest astronomical instruments of the period. It was divided into degrees and minutes to provide more accurate readings. The quadrant (pictured at right) had a radius of over six feet and faced an open window through which the observations were made. An observer, usually Tycho, and a timekeeper worked together to time the observations as precisely as possible. Then, a second assistant recorded the measurement called out by the timekeeper.

QVADRANS MVRALIS
SIVE TICHONICVS.

Upon its completion, Tycho's Uraniborg boasted the most state-of-the-art collection of astronomical instruments and observatories in the world. *(Tycho Brahe:* Astronomiæ instauratæ mecanica, *Wandsbek 1598. Courtesy of the Royal Library, Copenhagen.)*

Dirt from these excavations was used to form the square wall around the house. Gateways stood at the four points of the compass, and large dogs later resided in rooms above the east and west gates to announce the arrival of guests. At the north and south ends, small buildings in the style of the main house were built for servants' quarters. Under the servants' quarters were prison cells for troublesome tenants and servants. Inside the outer wall were orchards and flower and herb gardens. The square main house was constructed of red brick in the Gothic Renaissance style with each side measuring forty-nine feet. Each of the fourteen rooms contained a fireplace. The octagonal main tower was adorned with

CELESTIAL GLOBES are spherical maps of the constellations. A globe is mounted with poles that can be adjusted. The equator and the ecliptic (the Sun's path over the course of a year as viewed against the celestial sky) are drawn on the globe and divided into degrees, allowing astronomers to calculate the times of the stars' rising and setting.

Celestial globes have existed since ancient times. The oldest known celestial globe is the Farnese Atlas at the National Museum of Naples, Italy. The globe, a little over two feet in diameter and inscribed with the known constellations of the period, is actually part of an ancient sculpture of Atlas holding the heavens on his shoulders. This sculpture from the second century A.D. is a reproduction of a work from approximately 370 B.C., demonstrating how far back astronomical knowledge can be traced.

Over time, the globes became more detailed as astronomers gathered more information. The most accurate globes during the sixteenth century were produced with the data from Tycho Brahe's Uraniborg observatory.

In addition to Brahe's contribution to the accuracy of celestial globes, he is responsible for the manufacture of the globe at Uraniborg. Six feet in diameter and resting on a five foot high steel stand, this wooden globe covered in brass was perfectly spherical and maintained its stability despite the change of seasons. The positions of one thousand stars Tycho had charted were engraved on his globe.

Although the majority of Brahe's instruments were destroyed during an uprising of the Bohemians against the Hapsburgs, the globe of Uraniborg survived until 1728, when it was destroyed along with the Round Tower in Copenhagen that housed it.

a replica of Pegasus, the flying horse of Greek mythology, on its tip.

To make bricks, firewood was needed to feed the kilns, but there was little wood on the island. Instead, lumber was carried by boat across the sound from the forests around the royal manor of Kullagaard. In 1577, the king granted Kullagaard to Tycho, on the condition that he keep the lighthouse in running order.

Two towers at the north and south ends of Uraniborg had platforms with removable roofs for astronomical observations. The basement of the south tower contained an alchemy laboratory with sixteen furnaces. Tycho named a large circular room on the south end of the house his "museum." It served as a work station for his assistants, as well as a place to store valuable objects, such as his assortment of celestial globes. The museum eventually contained eight other celestial and terrestrial globes, several clocks, and a three thousand volume library.

The large southern observatory above the museum eventually housed a variety of instruments, including zodiacal and equatorial armillaries (instruments composed of rings to locate and demonstrate the position of celestial bodies), a vertical semicircle eight feet in diameter, a triquetrum (an instrument to measure the altitude above the horizon of heavenly bodies), a sextant with a radius of five feet, and a quadrant with a two-foot radius.

The large northern observatory contained a triquetrum

ARMILLARY SPHERE

Ancient in origin, the armillary sphere was used to measure and demonstrate positions and movements of heavenly bodies. These spheres were described by Ptolemy, but their use can be traced as far back as China in the first century B.C.

An armillary sphere consists of a set of rings, or armillaries. Each ring represents a "great circle" in the celestial sphere, the heavens, as seen from Earth. Some of the circles are fixed, and others revolve within the fixed rings.

An armillary sphere is used to plot the position of an object on the celestial sphere's surface. This mapping is done by referring to either the celestial equator or the celestial ecliptic. The celestial equator is the circle in the heavens that is directly above Earth's equator. The armillary sphere would be oriented by sighting on a celestial body whose position was already known. Once oriented, another ring with diametrically opposed sights is used to observe the celestial body in question. The position of this body was determined by the longitude from one ring and the latitude from another.

Tycho Brahe built many armillary spheres. Early on, he simplified the armillary sphere used by Ptolemy, which had five or six rings. Brahe's had four rings and the coordinates of the celestial objects could be determined directly without calculation. Brahe became dissatisfied, however, because the weight of the instrument was distorting the rings and affecting his data. As a result, he designed what is known as "The Great Equatorial Armillary Instrument, Comprising One and a Half Circles" for the Stjerneborg observatory. This enormous armillary was over eleven feet wide and rested on a sculpture of Atlas holding a globe. The one complete ring revolved, while the semicircle remained fixed.

Tycho was the last western astronomer to use armillary spheres for extensive astronomical observations.

sixteen feet in diameter, a sextant with a four-foot radius, a double arc to measure small distances, and a triquetrum that had been made and used by Copernicus. The great globe Tycho had ordered built in Augsburg several years earlier was installed in the library. His great mural quadrant was placed on a wall in the southwest room. It was solid brass with a radius of nearly seven feet. In all, Tycho's collection represented the largest, most accurate arrangement of astronomical instruments ever gathered in one place.

While Tycho and Kirsten were adjusting to their new life on Hven, they also began to have more children. Sadly, their first child, Kirsten, had died of the plague, but Magdalene turned three in 1576. A son, Claus, was born in January of the following year, but he lived only six days. Sophia, named after Tycho's sister and his paternal grandmother, was born in September 1578.

King Frederick liked and admired Tycho. In 1578, he granted the astronomer the use of eleven farms in the county of Helsingborg. Also, Tycho would be allowed to oversee the chapel of the Holy Three Kings, at Roskilde, as soon as the present canonry died. In the meantime, he could use income from an estate in Norway. The very next year, the overseer of the Holy Three Kings chapel died, and Tycho took over the canonry. The canonry at Roskilde included a residence, a mill, more than fifty farms, the right to appoint clergy at the two churches, and a country manor. His income from Roskilde was usually in the form of barley, rye, and oats.

Tycho did not always perform all his duties at Roskilde. It was traditional for the canon to regularly send money from the church's income to the widow of the previous canon and to the university. In December 1579, the king had to remind Tycho of this obligation. Tycho's income per year amounted to about twenty-four hundred dalers— about eight times what the highest paid professors at the university made, and about thirty times what the lowest-paid chief mathematicians made. In total, he controlled approximately one percent of the total wealth in Denmark.

For entertainment at Uraniborg, there was hunting and trapping of small animals, fishing in the stocked ponds, and games held in the orchards. From time to time, the king continued to hear from the peasants about harsh treatment handed down by Tycho. One repeated complaint was that Tycho misused his privilege of demanding peasant labor. Prior to Tycho's arrival, the peasants had been able to use all their time to work on their own farms. Now the new landlord demanded boon, or unpaid, labor, sometimes as often as four days a week. Others claimed that Tycho's bad temper caused him to overreact in tense situations. Finally, in 1581, King Frederick stepped in and set the amount of free labor per tenant to be no more than two days each week, sunrise to sunset.

CHAPTER FIVE
LIFE ON HVEN

Tycho would spend the next twenty-one years at Uraniborg, from 1576 until 1597. His main focus of study was astronomy, but he also pursued his interests in medicine and alchemy. It was also during these years that he finally rejected the heliocentric model of the universe and put forth his own planetary system. In all, these decades would be the most productive of his life.

Organizing and running Hven would have sapped the energy of most men. Tycho was able to turn the island into the most extensive and efficient scientific research center in Europe and also make steady progress, year after year, in observing and recording the sky.

The household at Hven was large and always shifting. Tycho adopted the ancient Roman idea of *familia* to govern Hven. Familia referred to all those who served at his will. The core of the *familia* was the household, including relatives, attendants, servants, and assistants. At Uraniborg, this included the basic staff—chaplain, tutors, wet nurses, a cook with helpers, bakers, a hops man, brewers, gardeners, a tailor, a button maker, watchmen, coachmen, fishermen, the housekeeper, maids, a secretary, and footmen. There was also a bailiff to direct the work of the household, an overseer to supervise peasant workers, a summoner to hail workers for specific tasks, clerks to keep records, and various skilled craftsmen, including masons, stone carvers, carpenters, tile workers, and a hydraulic engineer. The rest of Tycho's *familia* fanned out beyond the household to all peasants on Hven and to Tycho's patrons and supporters.

A colorful member of the *familia* was the court jester Jepp. Jepp was a dwarf who would sit at Tycho's feet during meals. Tycho treated him to morsels of food for his witty remarks. The dwarf was believed to have telepathic gifts. If a member of the household became ill, Jepp gave his forecast for recovery.

Once, two assistants whom Tycho had sent on a voyage did not return on time. Jepp said, "See how your people are laving [washing] themselves in the sea." When Tycho's men looked out into the ocean, they discovered that the assistants' boat had capsized, leaving both men drenched. At times when Tycho was away, the assistants used the opportunity to slack off on their work, and Jepp became their lookout. When he saw his master returning, he would yell, "The squire is on land!"

Tycho's wide travels and extensive education helped

him to expand his *familia*. He was fluent in Danish, German, and Latin; in addition, he knew Greek, some Hebrew, and probably spoke Dutch. This helped his *familia* become more cosmopolitan. Craftsmen came to Uraniborg from Wittenberg, Augsburg, Westphalia, Flanders, Holland, Norway, and Iceland, as well as Denmark.

Tycho and Kirsten's family also grew. In addition to Magdalene and Sophia, Elisabeth, named after another of Tycho's sisters and his maternal grandmother, was born in the summer of 1579. Tyge Jr., who Tycho hoped would follow in his footsteps and be an astronomer, was born in August 1581. Jorgen, named after Tycho's uncle, was born in 1583. The birth date of their last child, Cecily, is unknown.

Daily life in such a large household as Uraniborg required organization and the clear definition of roles. Nevertheless, it was often busy and chaotic, and privacy was limited. Servants rose at three or four o'clock in the morning and retired about eight o'clock in the evening. Cooks slept in the kitchen, Tycho's assistants in the attic, and servants in quarters located in the north lodge.

There were two meals served each day for all but boon (unpaid) workers, who were fed once. Beer was a safer substitute for water, which usually carried parasites, and subsequently, most residents consumed between four and eight quarts of beer each day. Breakfast consisted of warm beer and herring or rye porridge. The midday meal began at nine or ten o'clock with three or four courses. Diners used their own utensils, except during banquets, when knives and spoons were furnished. The chaplain opened each meal with a prayer. Tycho, his wife, and guests received their meals at the main table, while servants ate in the kitchen and boon workers ate at a grange outside the manor house.

Tycho needed educated assistants to help with observations, experiments, and record-keeping. There were usually between eight and twelve assistants working at Uraniborg at one time. These assistants received room and board, new shoes and clothing at Christmas, and other gifts.

Willem Janszoon Blaeu, a publisher of maps and scientific literature, spent six months visiting Hven. *(Artist unknown.)*

Tycho's sister Sophie was interested in alchemy, a popular science that was derived from Aristotle's propostion that all matter was formed from four essential ingredients. Based on this assumption, some medieval and Renaissance scholars believed they could find a way to turn ordinary metals into gold. *(Artist unknown. Courtesy of Gavnø Castle, Næstved, and Det Nationalhistoriske Museum på Frederiksborg, Hillerød.)*

Most of these assistants were in their early twenties and used the assistantship either as part of their university studies or as a break from school. They came from Denmark, Norway, Iceland, Holland, and Germany, and at least one came from England. The assistants averaged between 150 and 185 observation sessions a year. Usually the staff worked in teams of three, with the most experienced serving as the leader. One held the lantern and read out sightings, another called out the time, and the third entered the figures in a log. Tycho often assigned more than one team to make the same observation and would compare each group's records for inconsistencies.

Work for assistants went on from dusk to dawn, weather permitting. After a quick breakfast, the assistants were off to bed until early afternoon. They gathered again for a quick lunch, followed by a meeting where the night's work was planned and the previous night's observations were analyzed and recorded. At dusk, they began observing. The pace was hectic, and the master's temper could be fierce. Tycho had little patience for incompetence—or even honest mistakes. When he was displeased, his

TRIQUETRUM

A triquetrum is an instrument which predates Ptolemy (100-170 A.D.). It was designed to measure the angular altitude of stars using three straight rods and resembles a triangle. The rods are linked, with a vertical bar in the center and a hinged rod at its top and bottom. The two hinged rods can slide against each other, changing their point of intersection. One of these two rods is marked with graduations, while the other is fitted with sights. To determine the altitude of a celestial body, the vertical bar is held perpendicular to the earth, while one rod is used to sight the object of interest. The angle between the graduated and sighted arms, or the position of these two arms against each other, provides the altitude.

Tycho Brahe inherited a triquetrum, or Ptolemy's Ruler, from Nicholas Copernicus (1473-1543). A triquetrum could range in size. Some were hand held. Brahe, however, built a very large triquetrum, almost eleven feet long, at his observatory at Uraniborg.

roar traveled from room to room. Many assistants, who were usually young and homesick, found the pace and pressure too much and fled.

Part of Tycho's work included bettering the instruments he employed to make astronomical measurements. His modifications of the sextant made it more practical to use in astronomy, but he could not perform all the improvements on his own. He began to add craftsmen, sculptors, and architects to his staff at Uraniborg to design even more accurate and elaborate instruments. Among the instruments these craftsmen built was a six-foot steel quadrant and its brass arc with a six-foot radius. He also had an equally large double sextant built, which could be used to make observations simultaneously with the quadrant. When they were not working on instruments, the craftsmen built extensive gardens, intricate fountains, and even a system to provide running water throughout the home.

Soon, stories of Tycho's Uraniborg began to spread throughout Europe. People marveled upon hearing of his grandiose observatories with ceilings that magically parted to show the night sky, and his staggering collection of instruments—the largest anywhere. He became the most famous scientist in Europe—and one of the most envied—making it easier to attract top assistants. Many of Europe's best scientific minds spent time at Hven.

Twenty-three-year-old Peter Jakobsen Flemlose came to the island in 1578. Flemlose was a quick learner, and after Tycho taught him to use the cross staff and sextant, he gave Flemlose the job of assembling a new catalog of reference stars—fixed points—that could be used to pinpoint any moving object, such as a comet.

From April 1583 until 1597, Elias Olsen Morsing worked as one of Tycho's most favored and trusted assistants. Under Tycho's request, Morsing traveled with Flemlose to Frauenburg, where Copernicus had made his observations, to verify the latitude of Copernicus's observatory. While they were there, they were given the eight-foot-long triquetrum that Copernicus had made. They also received a self-portrait painted by Copernicus. They shopped for much-needed reference books before returning to Uraniborg.

Not all the visitors came to Uraniborg to merely assist Tycho. Sometimes, they hoped to gather information for their own purposes. Paul Wittich came from Germany during the summer and fall of 1580. In exchange for Tycho's information about observation methods, Wittich agreed to teach him a new method of trigonometry. After three months at Uraniborg, Wittich claimed that he had to go back home to help with the estate of a deceased uncle, promising to come back soon. He never returned, going instead to the neighboring estate of Landgrave Wilhelm IV at Hesse-Cassel and sharing secret information about Tycho's instruments, which Wilhelm used to construct his own.

One of the most important of Tycho's assistants was Christian Longomontanus, who worked with Tycho for eight years, from 1589 until shortly before Tycho's death in 1601. Longomontanus was born to peasant parents in 1562 in western Denmark. He helped his widowed mother run the family farm until entering a church school in 1577. After a year at the University of Copenhagen, he was recommended for an assistantship with Tycho. He became the best observer at Uraniborg—some said he was even more skillful than his master. He was also Tycho's personal secretary. Early in 1597, Tycho and Longomontanus began the enormous task of enlarging their star catalog that Peter Flemlose had worked on up to a thousand stars.

Tycho had still not decided which planetary system he accepted. Even with the new astronomical equipment, he had not been able to detect stellar parallax, which inclined him toward Ptolemy's Earth-centered universe. Before he made a final decision, however, he wanted to test another idea presented by Copernicus.

In *Revolutions of the Celestial Spheres*, Copernicus said that the distance between Mars and Earth is less than the distance between Earth and the Sun. This conflicted with what Ptolemy had written. If the distance between Mars and Earth could be measured, it might be possible to determine who was right.

Tycho began attempting to map the orbit of Mars. He designed a new mural quadrant that was sensitive and steady enough to make the observations. Consequently, over an extended period of time, most of Tycho's observations were focused on Mars. It was a task he never completed. After Tycho's death, Kepler continued the effort to map Mars's orbit by using these data—a project that eventually led to the discovery of his first two Laws of Planetary Motion.

It was during this time that Tycho finally decided to reject heliocentricism. Although his religious faith entered into the decision, most of his resistance to Copernicus continued to be based on physical mechanics. He was unable to accept the idea that Earth spins on its axis. In the next century, the Italian Galileo Galilei would perform a series of experiments that provided a

Longomontanus worked with Tycho for eight years, helping map the orbit of Mars and creating a star catalog. *(Courtesy of The Royal Library, Copenhagen.)*

mechanical framework for a revolving Earth. But Tycho was an astronomer, not a physicist, and he made his decision based on what he thought was the best empirical evidence.

Tycho did not devote all of his research time to studying the stars and planets. He continued to perform chemical research, which he called "terrestrial astronomy." Unlike the alchemists of the day, Tycho devoted little time trying to turn metals into gold. He was more interested in concocting remedies for the epidemic diseases that plagued the island, such as bubonic plague and typhus. Some of the ingredients he used were Venice treacle (drugs and herbs in a honey base), spirits of wine, aloes, myrrh, saffron, and sulfur. Tycho believed the solar phenomenon *aurora borealis*, or northern lights, was a result of a sulphurous vapor that altered the air's chemical composition. Therefore, he reasoned, earthly sulfur could cure diseases. He distributed his medicines to the island's inhabitants for free, and while some people saw this as an act of charity, others felt that he was using the peasants as guinea pigs.

His youngest sister, Sophia, the only member of his family who supported his desire to be an astronomer, spent a great deal of time at Uraniborg working as a research assistant along with the young men. Her particular interest was alchemy. Later in life, after Tycho's death, she married Erik Lange, a brilliant friend of Tycho's who dedicated his life to alchemy.

Although isolated at Hven, Tycho maintained friendships through letters and occasional visits. Two years after Tycho had visited him, Landgrave Wilhelm hired Christopher Rothmann as chief mathematician for his household, and Rothmann began to regularly correspond with Tycho. They discussed philosophical and technical matters. When Tycho learned that Rothmann was a Copernican, he tried to persuade him toward his own view that Earth did not move.

CHAPTER SIX
THE STAR WITH A TAIL

Toward the end of his life, Tycho wrote that his observations went through three distinct phases. First, during his time as a university student at Leipzig, was his "childish and doubtful" stage. Then there was his "juvenile and habitually mediocre" period, which covered the time between Leipzig and his arrival on Hven. Finally he entered his "virile, precise, and absolutely certain" time at Uraniborg.

Tycho's life did seem to be guided by the heavens. The solar eclipse he witnessed at fourteen set him on a career as an astronomer. The new star of 1572 that he wrote about in his first book convinced the king to grant him the use of Hven and the money to build Uraniborg. Then an accidental sighting on November 13, 1577, focused his energies for the next ten years and elicited his own planetary model.

Tycho was fishing in one of his ponds a little before sunset on November 13, 1577, hoping to catch his dinner, when he saw what appeared to be a star as bright as the planet Venus suddenly appear in the darkening sky. He rushed to the observatory, where he and his assistants took dozens of careful measurements over the next few days. He soon determined that it was not a star but a comet. He later learned it had been sighted in Peru on November 1 and in London on November 2. After the comet disappeared, Tycho set to work on a manuscript that was in various stages of composition and printing in the years from 1578 to 1587.

He began the book, eventually titled *Recent Phenomena in the Celestial World* but sometimes referred to as *Stella Caudata* (*The Star with a Tail*), by detailing his observations and measurements of the comet and its path. He measured the comet's tail and verified that it was always turned away from the Sun. At each stage, he presented his methodology as well as the raw measurements he used in making the calculations.

Tycho worked on *Stella Caudata* for over a decade. He and his assistants made thousands of measurements and included their data in the book, which became a model for later works of science. His chronicling of experiments and procedures used to arrive at conclusions meant that others, if they wanted to be taken seriously, would have to strive for similar clarity. It would no longer be enough to simply develop an impressive mathematical model.

Tycho's measurements proved that the comet originated and traveled well beyond the Moon, a phenomenon that directly opposed Aristotle's proposition that all change occurred in the sublunary sphere. In addition, Aristotle had written that all heavenly bodies moved in circular orbits, and that the planets were attached to a solid sphere. Tycho proved the comet's motion had not been circular and had intersected planetary orbits.

This work on the comet solidified Tycho's own ideas about the planetary system. In the last chapter of *Stella Caudata,* he explained that his "Tychonic system of the world" resulted from his efforts to determine whether Mars came closer to the Earth than the Sun did. Ptolemy had said the Sun was always closer; Copernicus had said Mars was closer. It took him several years to decide that Copernicus was right. This did not mean Tycho was willing to accept the idea of Earth rotating on its axis and revolving around the Sun, however. There were too many unanswered questions regarding motion and the size of the universe for him to make that leap.

Tycho said that he had little choice but to make a planetary system based on his own observations that also adhered to Aristotelian physics. In his system all the planets orbit the Sun, and the Sun orbits a stationary Earth. Because he had determined Mars came closer than the Sun did, its orbit intersected with the Sun's orbit. The stars, on the outer edge of the system, made one revolution every twenty-four hours and provided the energy that spun the planets, Sun, and Moon.

Tycho wrote about the comet of 1577, depicted in this woodcut as it passed over Prague, in his book *Stella Caudata*. The comet helped him to solidify his ideas about the structure of the universe. *(Courtesy of the Department of Prints and Drawings of the Zentralibibliothek, Zurich.)*

Tycho's system also allowed the orbits of Mars and the Sun to intersect, which implicitly rejected, once and for all, the Aristotelian idea that planets were attached to solid spheres. Planetary orbits were no longer physical roadways. Most importantly, it began the move away from Ptolemy's system, which was becoming increasingly harder to defend.

The historical importance of Tycho's system is that it was based on actual observations. It was designed only after years of studying planetary movements. Every aspect was dictated by empirical data. It was the first time that a serious, observing astronomer presented a cosmological model that, while it functioned mathematically, was restrained by the best evidence.

As he neared completion of *Stella Caudata*, Tycho became frustrated at his inability to find a reliable printer in Denmark. His manuscripts presented several difficulties. Sometimes new printing symbols had to be created to reflect the new developments in mathematics that were necessary to adequately illustrate his measurements. The new "science of triangles," or plane and spherical trigonometry, had to be duplicated on the page. There were elaborate illustrations and charts and careful attention had to be paid to continuity of text, graphs, and other figures. Plagiarism was always a concern, particularly to a man as jealous and possessive of his intellectual property as Tycho. This led him to establish elaborate security measures, which often created more difficulties.

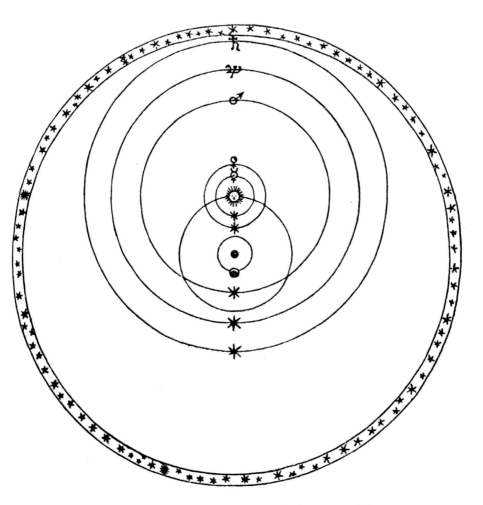

After rejecting the heliocentric model of Copernicus, Tycho proposed his own version of a geocentric system. In the Tychonic planetary scheme, all the planets orbit the Sun, which in turn orbits Earth. *(From Tycho Brahe:* Astronomiae Instauratae Mechanica, *1598. Courtesy of The Royal Library, Copenhagen.)*

Tycho hired the services of two outside printing presses, but a paper shortage in 1590 forced both presses to shut down. He then decided it was time to become a publisher. He had begun to build his own printing plant and support facilities on Hven in 1584. Building it was one thing, getting it operational, and keeping it running, was an even bigger job. He had to build a paper mill in order to have paper, which in turn needed swift running water to provide power. He also had to have access to forests to supply wood pulp for the paper. Tycho hired an assistant to build a series of dams and spillways to be used to power water wheels for a paper mill, a parchment mill, a gristmill, and a grindstone. When all the building was done, he hired a printer to publish his own works and those of others.

The first publications released by the Uraniborg press

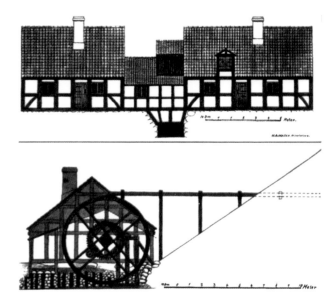

Tycho constructed a millhouse to produce paper for his bookmaking.

were a series of poems and woodcuts that paid tribute to Tycho's friends and patrons, and an astrological calendar. These publications helped to promote Uraniborg's scientific accomplishments. He also published an abridged, or abbreviated, version of *Stella Caudata*.

In the letter Tycho sent out with copies of *Stella Caudata*, he explained that he planned to expand the book and make it part of a three-volume culmination of his life's work. The first volume was to be about the new star of 1572; the second would explore the comet of 1577; and the third (which was never written) would deal with later comets. One of his goals in the work was to advocate empirical research, the discovery of scientific laws through observation. He also envisioned other works that would detail his instruments; the stellar catalog; his lunar, solar, and planetary motion theories; the latitudes of planets; and trigonometry.

Tycho did finish a more detailed book about the new star of 1572. Although he gave friends partial copies, it did not come out in final form during his lifetime. Its nine hundred pages consisted of three parts and a conclusion. In Part I, Tycho presented his solar and lunar theories and the observations for his catalog of stars. Part II described the star of 1572, as well as the instruments used to observe and measure the star, and stated the star's coordinates. Part III dealt with previous literature about the star. The conclusion explored the physical characteristics of the new star and its astrological significance.

During his twenty years at Hven, Tycho made regular astrological predictions for the king. When Frederick's heir, Prince Christian, was born in April 1577, Tycho attended the christening and prepared his horoscope. He predicted that the prince's infancy would be a safe one, with only one minor sickness. He warned that a serious illness when the prince turned twelve would not be fatal. When the prince turned twenty-nine, Tycho advised, he would have to carefully monitor both his health and his honor. Another crucial time would be in the prince's fifty-sixth year. Tycho said the royal prince would have a mild personality with some elements of seriousness. He forecast a life of mental alertness with interests in war, athletics, surgery, and science, but little interest in religion. The prince would not be fortunate in marriage, yet he would not be without happiness, especially in extramarital affairs with women. Considering how tumultuous the relationship between Tycho and the young Prince Christian would become in the near future, it is possible Tycho saw some things in his astrological chart he chose to keep to himself.

CHAPTER SEVEN
CALAMITY

Tycho dreamed that his island retreat would insulate him from the affairs of the royal court. In 1584, he built another observatory in a small hill about one hundred feet south of the first building. In order to protect his instruments from vibration and the wind, he had them installed underground. This observatory, fifty-seven feet square, contained five instrument rooms with a study in the center that held portraits of eight astronomers—Timocharis, Hipparchus, Ptolemy, Abattani, King Alphonso, Copernicus, Tycho himself, and Tychonides. This last painting was a portrait of his son Tyge, whom he hoped would follow in his father's footsteps and become a great astronomer. He called his underground observatory Stjerneborg, the castle of stars.

Despite Tycho's attempts at privacy, a steady stream of visitors interrupted his work. Professors and other

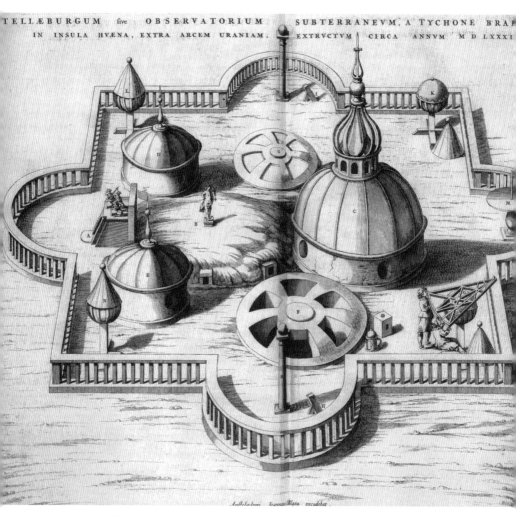

Tycho built an underground observatory inside a hill near his home on Hven that he named Stjerneborg, or castle of stars. *(From J. Blaeu:* Gran Atlas Cosmographie ou Blaviane, *1667.)*

intellectuals wanted to see the fabled Uraniborg and meet its legendary master. Some visitors were old friends. Tycho's former tutor and guardian, Andrew Sorenson Vedel, visited regularly. Twenty-eight-year-old Queen Sophia made a sojourn to the island in 1586. All the villagers came out to greet her, accompanied by musical fanfare. She toured the underground observatory as Tycho and his workers demonstrated the astronomical instruments. The queen was so taken with Uraniborg that two months later she returned with her father, mother, and cousin.

Uraniborg's reputation extended beyond Denmark. King James VI of Scotland, who later became King James I of England, visited in the spring of 1590, after marrying King Frederick's second daughter, Anne. During the festivities, James produced these lines of verse:

Muses' royal castle, jewel of the world, rivaling Olympus,
Nourishing house, your spirit's equal to your name
King James composed these lines in honor of Tycho:
What Phaethon dared was by Apollo done,
Who ruled the fiery horses of the sun.
More Tycho doth, he rules the stars above,
And is Urania's favorite, and love.

Still, some visitors were more interested in laying their hands on Tycho's research than simply enjoying his hospitality. In 1584, Nicholas Reymers Bar, who called himself Ursus (the Bear) visited Uraniborg. Tycho thought that Ursus was overly curious about the instruments and books and became suspicious of his guest. He ordered an assistant to steal into Ursus's room and go through his pockets. The servant found notes about Uraniborg. When Ursus discovered that his papers had been taken, he became enraged. He insisted they had nothing to do with Tycho's research. Tycho returned the papers, but the incident prompted him to tighten security at Uraniborg. He began requiring that his assistants sign contracts with him that spelled out their rights, duties, and responsibilities.

Frequent visitors were not the only sources of distraction for Tycho. As Denmark became more powerful politically, a need for more maps grew, and the king enlisted Uraniborg to print them. Also, with the birth of each of the king's sons, Tycho had to create a horoscope. The king's interest in weather led to Tycho's compiling a book on the subject.

As Uraniborg grew, Tycho worried about who would take over after his death. His sons could not inherit it because the law stated that any sons born of a common-law marriage between a nobleman and a common woman could not inherit their father's property. There was one loophole. A nobleman could give his personal property to his commoner children while he was still alive, and upon his death, the children could keep it.

In order to grant Hven to his male heirs, Tycho would need the king's support. With Vedel's help, he wrote a

Queen Sophia of Denmark was one of many visitors who were impressed by Tycho's Uraniborg. *(Courtesy of The Royal Library, Copenhagen.)*

91 ✻

ICHNOGRAPHIA STELLÆBURGI

As Tycho's popularity grew, he hosted many visitors who wished to tour Uraniborg. This plan of Stjerneborg, his famous underground observatory, is taken from his book, *Astronomiæ instauratæ mecanica. (Courtesy of The Royal Library, Copenhagen.)*

document for Frederick to sign. Uraniborg would be considered an institution of higher education, as commoners could legally supervise a school. Sometime between 1582 and 1588, Tycho presented his plan to King Frederick and got his verbal approval; however, he failed to get the document signed before the king died in 1588.

King Frederick II was a religious, gracious, and open-minded ruler, as well as a generous patron to Tycho. He did, however, have one fault that Vedel, who had become a prominent writer and historian, pointed out during the oration he delivered at Frederick's funeral. He wrote, "If his grace could have kept from that injurious drink which is much too prevalent all over the world among princes and nobles and common people, then it would seem to human eyes and understanding that he might have lived for many years to come."

When Frederick died, his son Prince Christian was only ten. Queen Sophia tried to claim the right to govern the country until her son came of age, but the nobles, jealous of the power she would have, decided that four representatives from the privy council would act as a regency council to govern the country for the next ten years. Tycho, anxious to see how much support he had on the regency council, asked for an additional six thousand dalers per year. When the money was granted, Tycho assumed that he would be able to continue working at Hven with state support. In August 1588, the regency council also approved Tycho's plan to have one of his sons inherit Uraniborg.

Tycho continued on state projects in order to remain in good stead with the regency council. Vedel, who was serving as the official historiographer of Denmark, needed a new map of the country, and Tycho agreed to help. One of his assistants traveled with Vedel to determine the latitudes of various locations. Then Tycho and Vedel performed the first survey in history based on triangulation, a method of measurement that uses triangles to determine the distances between certain points.

In March 1590, Tycho received a letter from Queen Sophia ensuring that she had heard King Frederick say he would appoint one of Tycho's children as his successor to administer Uraniborg. Tycho kept the letter to help protect him from future trouble.

Although Tycho seemed to have the support he needed at court, all was not well on Hven. In 1589, the court jester Jepp tried to escape. He was captured and flogged. The next year, two servants were expelled after they attacked a young girl. Tycho jailed his tailor for three days for an unknown indiscretion, but the tailor escaped from the island and complained of his mistreatment.

In a 1591 letter, Tycho complained to Landgrave Wilhelm about "unpleasant obstacles" that interfered with his work. He alluded to the possibility of relocating. The letter was prompted because of a long and bitter dispute he was having with a tenant named Pedersen, who leased from Tycho a small manor, consisting of nine cottages and a manor house. When the manor house was destroyed in a fire, Pedersen had it rebuilt, and moved the cottages to extend the size of his fields. He also built additional cottages. After he completed the improvements, Tycho tried to raise Pedersen's rent. When Pedersen refused to pay, Tycho evicted him in October 1590, refusing to refund his lease fee. Tycho ordered his workers to plow and sow the fields; Pedersen's men sowed rye behind them in protest, thereby ruining the crops. Tycho then had Pedersen put in irons and taken to Hven to be held as a prisoner for six weeks. He also had all of Pedersen's business records confiscated. The case ended up in a local court, where it dragged on for months. Finally, the court decided in Pedersen's favor. The publicity around the trial damaged Tycho's reputation both on Hven and at the royal court.

In 1591, Landgrave Wilhelm IV requested that Tycho send him an elk from Hven for his private zoo. The group delivering the animal stopped one night at the home of a relative of Tycho's. The elk slipped into the manor house, drank some beer, fell down the steps on its way back out, broke a leg, and died. In a letter explaining the mishap to Wilhelm IV, Tycho alluded to the possibility of moving to another country. Privately, he wondered if the mishap could be an omen of things to come.

In July 1592, fifteen-year-old Prince Christian visited Hven. Upon the prince's arrival, servants washed his hands and welcomed him with toasts. Tycho took the prince on a tour of the property. The weather was beautiful and the future king enjoyed himself. His favorite part of Uraniborg was the planetarium in the museum. Christian liked a small brass globe that showed the motions of the Sun and the Moon so much that Tycho gave it to him. In return, the prince gave Tycho a gold chain with his portrait attached.

During the visit, the two discussed military tactics, shipbuilding, astronomy, and navigation. Christian was especially interested in constructing a school on Hven, but Tycho knew it would take away from his own work. Nevertheless, Tycho was awarded an annual grant of four hundred dalers to teach young people the arts of navigation and astronomy, and another 120 dalers per year to pay for the students' room and board. Tycho accepted the assignment in order to remain in the good graces of the crown.

Soon after Christian IV was crowned king of Denmark in 1596, Tycho would lose the patronage of the royal family that he had enjoyed under Frederick II. *(Courtesy of The Royal Library, Copenhagen.)*

Christian's goodwill was short lived. The following summer, Prince Christian traveled to the Chapel of Holy Three Kings in Roskilde Cathedral, which Tycho had been granted supervision of years before, and saw that the chapel where his father lay buried was in disrepair. The regency council ordered Tycho to make the repairs by winter, or it would be removed from Tycho's supervision. Tycho had the work completed in time, but the incident had lost him the support of several of his allies at court. Two of his other supporters died over the next two years.

Slowly, the climate at the royal court changed toward Tycho. During Tycho's youth, the privy council, made up of noble friends and relatives of the Brahes, had undertaken major decisions and prevailed upon the king to carry them out. However, with the establishment of the regency council after King Frederick's death, enemies of the Brahe family became more outspoken. As the prince approached the age when he would become king in his own right, this group began to gain Christian's trust. Tycho's brother, Jorgen, lost his post as regional governor, a cousin was ousted from a castle, and his brother Steen lost three fiefs.

Christian was young, full of energy, and increasingly dedicated to changing things. Tycho was forty-eight years old and growing tired. He complained that the frequent, lengthy commutes between Hven and the mainland were exhausting. He had not been out of Denmark for almost twenty years and was restless. He missed closer contact with intellectuals. It was probably a combination of these pressures that prompted him, in August 1594, to sell his interest in the family castle Knudstrup to his brother Steen. Perhaps he was also planning ahead: He could legally transfer cash to his sons, but not property, at his death.

During the six years that he was at Uraniborg, Gellius Saescerides, one of Tycho's assistants, became interested in Tycho's eldest daughter, Magdalene. Thinking them a good match, and hoping to prolong the thirty-two-year-old assistant's tenure, Tycho agreed to the engagement. Trouble soon ensued. Gellius angered Tycho by greedily demanding a larger dowry (the money the bride's family was required to give the groom). Tycho refused and demanded that Gellius work at Uraniborg for a year after the wedding. This arrangement would provide a better standard of living for his daughter and would free Tycho to complete more work. However, the prospective groom had plans to become a professor of medicine at the University of Copenhagen and did not want to stay on Tycho's island.

When Tycho was informed that Gellius intended to scale down the size of the wedding even more, he became angry, but eventually agreed. Then Gellius broke off the engagement, which may have been his goal all along.

Magdalene was not upset by the breakup. She even wrote, "How happy I am that the lad has so graciously freed and delivered me!" But Gellius began to spread rumors that Tycho thought tarnished his family's reputation. Tycho sued him and succeeded in clearing his family's name, but he had lost a gifted member of his *familia* and made public another conflict with an assistant. Magdalene would remain unmarried for the rest of her life.

Meanwhile, the travails at Hven continued. In March 1596, a servant committed suicide, and in May, a jester named Per Gek was brought back in chains after trying to escape. The following month, the jester and a student named Jacob succeeded in escaping. A dead sailor was found buried on the beach, and in July, Tycho dismissed three members of the *familia* for using the laboratory without his permission.

CHAPTER EIGHT
STARTING OVER

Nineteen-year-old Prince Christian IV was crowned king August 29, 1596, the year Tycho turned fifty. In an effort to economize, the new king suspended Tycho's Norwegian fief immediately after the coronation. Meanwhile, enemies of the Brahe family kept the king informed about every complaint lodged against Tycho. One of his enemies was the king's physician, Peder Severinus, who may have been jealous of Tycho's tendency to distribute medicine without demanding payment, thereby cutting into Severinus's profits. Another enemy, Christopher Valkendorf, was an advisor and treasurer to the king. It was rumored that his trouble with Tycho started when the owner of two dogs, who had promised to give one to each man, sent the better dog to Valkendorf. This angered Tycho, who let his displeasure be known. The new chancellor Christian Friis was also openly hostile to Tycho. Finally, in March of 1597, Tycho's annual salary of five hundred dalers was stopped. In effect, the great Tycho Brahe had been fired.

The conflict between Tycho and King Christian went beyond a mere personality clash; their philosophical differences stood behind the king's decision to cut Tycho off. Whereas King Frederick had favored research over education, the new king favored education over research. The result was that university officials, who had for years watched their research projects go to Uraniborg, now watched with relief as their funds returned to the university. King Christian's behavior was partly prompted by his religious beliefs. King Frederick and Tycho had agreed that through man's free will and intellect, God's secrets could be unveiled. King Christian and his advisors believed that science and religion were separate entities.

King Christian also represented a new philosophy that was becoming prevalent among rulers in Europe. Whereas King Frederick's actions had been checked by the powerful nobility, King Christian was determined to limit the nobles' power. He became an advocate of a new doctrine called "the divine right of kings," proposing that kings ruled as direct agents of God.

Once the residents on Hven knew Tycho was no longer a favorite of the court, they complained even more about his oppression and ill-treatment of them. When the king heard these complaints, he ordered an investigation led by Tycho's rival Chancellor Friis. At a hearing, Friis questioned peasants in the presence of King Christian and Tycho. The peasants hinted that Tycho had pocketed money meant to be used to maintain the island church, that he had expropriated church lands, torn down the parsonage buildings, fed the pastor along with common laborers, and dismissed pastors arbitrarily. Finally, he had allowed the ritual of exorcism to be eliminated from the baptism ceremony, a theological omission that offended the islanders' Lutheran beliefs.

Losing the support of the crown devastated Tycho's ego and prestige. It was almost more than he could bear. He relocated, in typical grand fashion, with his family, assistants, and servants, to his house in Copenhagen. He took along his printing press, library, furniture, and all observational instruments except the four largest ones.

The only charge brought against him that officials were able to verify—because it was inarguably true— struck directly at Tycho's marriage and family. Reverend Wensosil, a minister on Hven, "was dismissed in disgrace for not having . . . punished and admonished Tycho Brahe of Hven, who for eighteen years had not been to the Sacrament, but lived in an evil manner with a concubine." In other words, the minister was fired for not denouncing the relationship of Tycho and Kirsten, who were not legally married. The pastor was imprisoned for a month, but the real target of the charge was Tycho; the legitimacy of his entire family was called into question. It also meant the king had no intention of honoring Tycho's wish to leave Hven to his sons.

After Tycho became lord of Hven, he began to have problems ruling the island. *(Courtesy of Gavnø Castle, Denmark.)*

…IGIES TYCHONIS BRAHE OTTONIDIS ÆTATIS SVÆ

In June, with no more work to do, or money to pay them, Tycho let his assistants go. He attempted to carry out observations by himself on a city wall near his home in Copenhagen, but the city forbade him. The local government may have had an ordinance against making observations from the wall, but it is more likely that his enemies in court, who now included the king, were creating roadblocks to stop Tycho from reestablishing his observatory on the mainland. The government also forbade Tycho from carrying out chemical experiments in his home. Portraits of Tycho painted during this period reveal a patriarch with thin hair, drooping mustache, and weary eyes.

After three months in Copenhagen, Tycho wrote Duke Ulrich of Mecklenburg, asking him if he and his family could stay temporarily in Rostock, Germany. When the duke agreed to his request, Tycho announced that he was choosing "exile over dishonor," and sailed with his family to Rostock. In addition to servants, fifty-year-old Tycho was accompanied by Kirsten; their six children, ranging in age from fourteen to twenty-three; and the deposed pastor from Hven, Reverend Wensosil. In all, more than twenty people traveled to Rostock, where they were greeted warmly by friends.

While Tycho waited in Rostock, his one goal became to reestablish a domain to publish his works on astronomy and provide for his family's security, whether in Denmark or elsewhere. He wrote again to King Christian IV, asking him to reconsider. He reminded Christian of the support his father had given him, including his promise that Tycho's sons could inherit Uraniborg after his death. Since Christian had become king, however, Tycho lamented that everything had changed:

> But it has turned out differently from what I had believed, about which I shall now only state the following. Your Majesty is doubtless aware that I have been deprived of what I should have had for the maintenance of the said [astronomical] art, and that I have been notified that your Majesty does not intend further to support it, in addition to much else which has happened to me (as I think) without my fault or error. And whereas I, by the grace of God, shall have to carry to an end what I once with so much earnest and for so long have worked at, which is also known to many foreign nations and greatly desired, and I have not myself means for this, as I have been so reduced that I, notwithstanding the fiefs I held, have been obliged to part with my hereditary estate.

He asked the king to forgive him for leaving the country in search of a new patron. He said he would return to Denmark "if it could be done on fair conditions, and without injury to myself." He also asked sympathy:

> It is by no means from any fickleness that I now leave my native land and relations and friends, particularly at my age, being more than fifty years old and burdened with a not inconsiderable household, which I, at great inconvenience, am obliged to take abroad. And that which is still left at Hven proves that it was not

formerly my purpose and intention to depart from thence.

As the exiled astronomer waited for a reply from King Christian, he may have wondered if he had made a mistake by leaving Denmark so quickly. Had Tycho stayed, he might have been able to smooth things over enough to keep his domain. His haste had been prompted by the experience of his friend Vedel, who had been removed as the court's historiographer after all of his research was confiscated. Tycho had moved quickly for fear that the same thing might happen to him.

While putting pressure on King Christian to reinstate him, Tycho also looked for a patron outside his homeland. Heinrich Rantzau invited him to stay temporarily in one of his castles near Hamburg, Germany. Tycho chose Wandsburg, located three miles northeast of the city. In September 1597, the entourage arrived in a fancy coach drawn by six horses. Once there, he met with some former colleagues and worked on an illustrated description of his instruments. In February 1598, he arranged to have his larger instruments sent to him.

Tycho finally received King Christian's cold response to his letter. "Your letter is somewhat peculiarly styled and not without great audacity and want of sense, as if we were to account to you why and for what reason we made any change about the crown estates." The king did hint that if Tycho asked nicely, he might win back the title of chief mathematician.

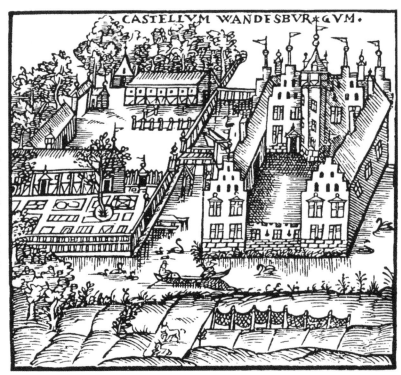

After leaving Hven, Tycho and his family lived for a short time in the castle of Wandsburg. *(Woodcut by unknown artist, 1590.)*

In 1598, Tycho finished a book on his instruments. In an attempt to win a new patron, he dedicated the book to Rudolph II, emperor of the Holy Roman Empire, a loose confederation of mostly German states in central Europe. Rudolph had moved the royal court from Vienna to Prague in the state of Bohemia. Tycho also included a new star catalog as a New Year's gift to the emperor, along with a letter asking that he be allowed to continue his work in Germany. Although Tycho did not send a portrait of himself in his book to Rudolph, in copies he presented to friends he had pasted a large watercolor of himself looking bald, drawn, and sad.

CHAPTER NINE
ENTER KEPLER

Soon after leaving Denmark, Tycho became embroiled in another quarrel, this time regarding his work and other astronomers, that had been brewing for almost ten years. In 1588, after receiving his manuscript copy of *Stella Caudata*, Christopher Rothmann, chief mathematician for Landgrave Wilhelm IV, wrote back to Tycho and compared it to another book he had read that proposed a very similar planetary system. This puzzled Tycho, who did not recognize the author's name of the book Rothmann had read. Later, Tycho found a copy of the book in question and discovered it had been written by Nicholas Reymers Bar, also known as Ursus (the Bear), the man who had visited Hven and was discovered to have notes about Uraniborg in his pockets. The planetary system in his book, *Foundation of Astronomy*, differed from Tycho's in only a few points.

Tycho became enraged. He wrote to astronomers all over Europe informing them that Ursus had stolen his ideas. One of the points in which Ursus deviated from Tycho was that he allowed Mars to circle the Sun without ever intersecting the Sun's orbit. Tycho argued that Ursus could "have learned that this is impossible even from a superficial glance at Copernicus." Clearly, he said, one who "has not mastered the precise observations of many years so as to elicit adroitly from them the defects in the hypotheses . . . is pitiful and thoroughly ludicrous. For he lacks the means by which to prove from the phenomena why the earlier views must be abandoned and revised in this way, and how to establish by manifold observations that this innovation conforms more closely to the heavenly phenomena."

Thus began a long and bitter feud that became infamous within Europe's scientific community. Ursus began to attack Tycho personally, ridiculing his nose and slandering his common-law wife. The case against Ursus was eventually strengthened when he was charged with plagiarism in a separate incident. He had forecasted the death of Emperor Rudolph's young brother, but the prediction was discovered to have come from Elias of Prussia. For this transgression, he lost his job as imperial mathematician at the court in Prague, his book attacking Tycho was banned, and the mathematician left Prague in disgrace. This outcome did little to relieve Tycho's obsession with revenge, however, and Ursus became Tycho's scapegoat for all of the frustrations he had suffered in the last years.

In the middle of this feud with Ursus, Tycho received a copy of a recently published book by Johannes Kepler, a young mathematics professor at a seminary in Graz, in present day Austria. The book, *Cosmological Mystery,* was one of the first books written by a mathematician and astronomer who supported Copernicus's theories. By a stroke of bad luck, which seemed to be Kepler's constant companion, the book arrived in the mail the very same day that Tycho received his copy of Ursus's slanderous book.

Cosmological Mystery came accompanied by a letter of introduction from Kepler flattering Tycho and begging "his greatness" to read and critique his "feeble, inept, and childish" words. Tycho read the book and responded that he admired Kepler's prose style and the quality of his "speculation," but rejected the Copernican thesis with an explanation of why his planetary system was superior. He finished the letter by inviting Kepler to visit him in Germany.

Below his signature, Tycho added a note that must have struck fear in Kepler's heart:

> I do not know by what chance it happened that the same courier who brought me your letter . . . at the same time also delivered to me . . . a defamatory and criminal publication, tied together with your letter, and written by a certain wild Ursus . . . I could run through the insolent chicaneries and shocking lies and insults with which his publication abounds to excess . . . Yet I also found there a certain brief letter by you

IOANNIS KEPPLERI
Mathematici Cæfarei
hanc Imaginem
ARGENTORATENSI BIBLIOTHECÆ
Confecr.

MATTHIAS BERNEGGERUS
b al. Ianuar. Anno Chr.

M DC. XXVII

Johannes Kepler would continue some of Tycho's projects after the astronomer's death, including his mapping of the orbit of Mars and the Rudolphine Tables. Most important, however, would be Kepler's three laws of planetary motion, which would lay the groundwork for further study in physical astronomy. *(Erich Lessing/Art Resource, NY.)*

[printed inside the book] with which he strives to adorn himself and conceal his misdeeds. When I read your letter, I was astonished that you made so much of him.

Tycho continued to explain that everything Ursus had published had been "snatched and stolen from others, as is his habit." While he forgave Kepler for his praise of Ursus, he did wish that Kepler had not "done so to excess!"

After torturing Kepler with his displeasure, Tycho finally let him off the hook. "But you wrote these things through ignorance when you were young." Tycho's forgiveness was conditional, however. He asked Kepler to put in writing a retraction of his praise of Ursus, which Tycho planned to use in a court battle against Ursus.

Tycho's response was a remarkably restrained reaction, considering his infamous temper. His restraint was probably due to the fact that Tycho knew the young Kepler, who was already respected as one of the most gifted mathematicians in Europe, was about to lose his job because of religious persecution. He also knew that Kepler was desperate to find a new position. Tycho no longer had a team of assistants, and he wanted Kepler to come work for him. He thought the young German might be the man to carry on his work and ensure his legacy by validating the Tychonic system.

Johannes Kepler was twenty-eight when he first began corresponding with Tycho. He had been a student at the University of Tübingen under the mathematician Michael

Tycho accused Nicholas Reimers Ursus of plagiarism after Ursus published his *Foundation of Astronomy. (Strasbourg, 1588.)*

Maestlin, where he first learned about Copernicus. After earning his master's degree in 1591, he entered theological seminary, but it soon became clear to his teachers that he was more suited for a career in mathematics and astronomy.

Kepler was offered a position as a mathematics professor at Graz but hesitated before taking it. He had made some enemies at Tübingen and suspected the appointment had been arranged as a way to get rid of him. Also, he was a Lutheran, and Graz was under the control of Archduke Ferdinand, who was committed to returning all Christians, by the use of force if necessary, to Catholicism. It could be an uncomfortable, maybe even dangerous, place for him to live. Worse yet, the school was small and not well known. Even with these doubts, Kepler soon realized he had no choice but to take the job. In Graz he was supposed to teach mathematics, but he spent most of his time lecturing in classics and rhetoric because few students signed up for mathematics. He also prepared astrological calendars and individual charts, including ones for himself and family that still exist today, providing a remarkable historical record.

Kepler published *Cosmological Mystery* in 1596. Soon after its publication, Archduke Ferdinand closed the school in Graz for several months. When he re-opened the school, he replaced all the Lutheran teachers with Catholics, decreeing all Lutherans had to leave in eight days. His one exception was Kepler, whose brilliance had brought prestige to the school. Kepler could stay in Graz for the time being, but he would not be permitted to continue teaching unless he converted to Catholicism. Kepler refused.

Kepler's misfortune was a stroke of good luck for the future of astronomy. With his job possibilities dwindling, he turned to the one man who could help. Kepler wrote to Tycho, apologizing for his letter to Ursus. He hoped this apology would be enough to convince Tycho to offer him a position. In December, Tycho wrote that he wanted to meet Kepler soon, assuring him that he would be a loyal friend. The letter was enough for Kepler to remain hopeful for a change in his fortune.

CHAPTER TEN
REBIRTH

Tycho was able to hire Kepler as his assistant because of his prosperous relationship with Emperor Rudolph. The emperor, whom Tycho hoped to gain as a patron, was more interested in science and art than politics and governing. He was a member of the Hapsburg dynasty that ruled the Holy Roman Empire for over four centuries. The empire consisted of a loose federation of smaller states in Central Europe, including most of present-day Germany and Austria. Rudolph was a Catholic but refused to join in the Catholic effort to stamp out the "Protestant heresy." His goal was to make Prague a center of research and education. He succeeded for a while, but the impending religious conflict would spread through most of his empire, forcing his abdication in 1611, and culminating in thirty years of highly destructive war.

Tycho and Rudolph's meeting had been arranged for January 1599, but because of bad weather, it was May of that year before Tycho packed up all his belongings and left Wittenberg. He arrived in Prague at the beginning of July bearing several letters of recommendation, a copy of the star catalog he had prepared for Rudolph, an illustrated catalog of his instruments, and a presentation copy of his theory of the solar system.

Tycho and one of his sons rode to the emperor's residence, Hradcany Castle. In an account shortly after the interview, Tycho told a friend what had transpired:

> I went in to the emperor . . . and saw him sitting completely alone in the whole room, without even an attending page . . . The emperor immediately called me over to him with a nod, and when I approached him, he graciously reached out his hand to me . . . I then drew back a bit and made a short speech in Latin . . . He responded graciously to me with a more detailed speech than mine, saying, among other things, how agreeable my arrival was to him, and he promised to support me and my research, all the while smiling in the most kindly way, so that his whole face beamed with benevolence. I humbly thanked him for this proof of his grace and mentioned the three books I had with me to present to him with the utmost deference. When he graciously responded that he would accept them, I immediately fetched them from my son . . . When he took them and laid them out on the table, I reviewed the contents of each briefly. Then the emperor again responded with a splendid speech, saying most graciously that they would please him greatly.

Emperor Rudolph agreed to become Tycho's patron after the astronomer moved to Prague. *(Courtesy of The Royal Library, Copenhagen.)*

Emperor Rudolph offered Tycho a position with a salary of three thousand florins a year (Kepler had made two hundred florins a year teaching in Graz). There would be additional income from other sources, and he promised Tycho a hereditary estate, which would provide security for his family after his death. As for living quarters, the emperor offered a choice among three castles: Lyssa, Brandeis, and Benátky. Tycho chose Benátky, situated on the River Iser, a tributary of the Elbe. Twenty-two miles northeast of Prague (six hours away by level road), Benátky had been built on a hill two

hundred feet above the river in 1522. It looked out over acres of beautiful vineyards, orchards, fields, and pastures.

Tycho immediately began to renovate the castle to install his observatories and laboratories. He sent his eldest son, Tyge, to Denmark with Claus Mule to arrange for the removal of the four large instruments from Hven. The instruments arrived in November.

Kepler began working for Tycho in February 1600 at Benátky Castle. Tycho quickly sized up Kepler's brilliance and mathematical genius. Tycho set up a division of labor, as he had at Uraniborg. After Tycho asked him to sign a pledge of secrecy, Kepler would devote his time to studying planets, especially Mars. Tycho's younger son, Jorgen, was in charge of the chemical laboratory. His longtime assistant Longomontanus would make lunar observations.

Kepler was most interested in mapping the orbit of Mars. None of the systems to that point, including Tycho's, had satisfactorily solved the problem presented by the planet's eccentric orbit. Kepler was convinced that the only solution was to map the entire orbit. Hopefully, this would enable him to determine exactly where the change in orbital speed took place.

From the beginning of their relationship, the two men established a pattern of estrangement and reconciliation. They were different in many ways. Tycho was fifty-three; Kepler was twenty-nine. One was a nobleman, used to ruling others; the other a commoner who preferred books to people. Tycho had not received the sacrament for eighteen years; Kepler was devout and had planned to enter the clergy. Tycho, despite his volcanic temper and suspicious nature, was warmhearted and extroverted; Kepler was shy, occasionally petulant, and introverted. Kepler's frustrations often arose because every morsel of data had to be pulled from Tycho. In a letter, he complained of feeling more like Tycho's assistant than his equal.

What held them together, despite their differences, was a mutual commitment to astronomy. Although he rejected the Tychonic system, Kepler appreciated the fact that it had been developed by a working astronomer. He had little faith in Tycho as a theoretical scientist, but boundless respect for him as an accumulator of accurate measurements. As Kepler wrote in a letter to his former teacher Maestlin:

> Tycho possesses the best observations and consequently, as it were, the material for the erection of a new structure; he has also workers and everything else which one might desire. He lacks only the architect who uses all this according to a plan. For, even though he also possesses a rather happy talent and true architectural ability, still he is hindered by the diversity of the phenomena as well as by the fact that the truth lies hidden extremely deep within them. Now old age steals upon him, weakening his intellect and other faculties, or, after a few years, will so weaken them that it will be difficult for him to accomplish everything alone.

It is more difficult to determine exactly how Tycho felt about Kepler. Tycho was, after all, in the superior position. Maybe the best evidence is that after every blow up, which usually ended with Kepler flinging insults and fleeing Benátky, Tycho welcomed him back when the chagrined Kepler asked to return.

Their working arrangement at Benátky would be short-lived. At the emperor's request, after only a few months, Tycho and his family left the castle and moved to a house Rudolph had purchased for him in Prague. Kepler and his family would later join him, but first Kepler wanted to try to settle old business at Graz. He wrote Tycho that because he had refused to convert to Catholicism, he would not be granted a leave of absence from the seminary, and therefore he would not receive a salary from his old job. He asked Tycho to intervene with the emperor and arrange for him to be paid a large enough salary to support his family.

Fortunately for Kepler, Longomontanus had recently returned to the University of Copenhagen to take a job as a professor. Tycho needed Kepler's help more than ever. There was still tension over money, as there always was with Tycho, but he had no choice but to ask the emperor to place Kepler on the payroll. When Tycho and his wife and children were settled in his Prague residence, there was space provided for Kepler and his family.

Tycho desperately wanted to publish his long overdue books. Earlier versions had been limited to gift copies for friends and potential sponsors. As he turned fifty-four, making new observations was no longer a priority. Instead, he wanted to concentrate on organizing his earlier work into manuscripts. Their publication, he hoped, would establish his place in history. He tended to talk of the past rather than plan new projects. The autobiographical part of his writings often sidetracked him away from writing about science, as did his continued preoccupation with Ursus, who he saw as a threat to his immortality.

There was another factor distracting Tycho—he was homesick. Kepler noted that he spent a great deal of time writing letters to Denmark. In one two-day period, he wrote nine letters, bragging about his accomplishments. Behind the bravado was an obvious longing and loneliness. The only person from his birth family who had remained close to him was his sister Sophia, who no longer lived with her husband. He looked forward to making a trip to Denmark during the summer of 1601 to see relatives and friends and to attend the annual meeting of the Danish nobility.

Meanwhile, in April 1600, Frederick Rosenkrantz, a third cousin who was also in exile, visited Tycho. Rosenkrantz was a colorful character. He had been raised as a nobleman, but was forced to flee Denmark after he had had an affair with a lady-in-waiting to the queen. Authorities captured him and brought him back for his punishment—the loss of two fingers and his nobility—but the sentence was reduced to government

English playwright William Shakespeare might have named a character in his tragedy *Hamlet* after Frederick Rosenkrantz, Tycho's third cousin. (*National Portrait Gallery, London. Portrait by John Taylor, 1610*)

service in a war against the Turks. While he was in London with a Danish delegation in 1592, he impressed the twenty-eight-year-old William Shakespeare and was granted a bit part in *Hamlet*. One of the tragedy's minor characters was named Rosenkrantz.

Another visitor to Prague in June 1600 was David ben Solomon Gans, who wrote a description of Tycho's impressive observatories—thirteen rooms in a row, each with a large instrument, and twelve assistants at work. He did not know that this was a much less impressive display than had existed in Uraniborg.

An intermittent fever and bad cough hindered Kepler's work. He also angered Tycho by stubbornly clinging to the heliocentric model. In April 1601, Kepler went back again to Graz. Upon his return in August, the emperor promised him the office of imperial mathematician if he would work with Tycho on the new planetary tables that had been promised. Tycho had asked permission to name the work the *Rudolphine Tables*, and the emperor had agreed. This was exciting news to Kepler; it meant that he was in line to be Tycho's successor. However, once he returned to work, much of his time was spent refuting the work of Ursus in preparation for a legal battle. Fortunately for Kepler, Ursus died before the trial began, and he was finally free to work on the tables and his own research on the orbit of Mars.

CHAPTER ELEVEN
END OF AN ERA

On October 13, 1601, Tycho was invited to dine at the Baron of Rosenberg's house. During the meal, he felt the urge to relieve himself, but he did not leave the table for fear of offending his host. Kepler later recounted what Brahe told him of that evening: "Holding his urine longer than he was accustomed to doing, Brahe remained seated. Although he drank a bit over generously and felt pressure on his bladder, he had less concern for the state of his health than for [the breach of] etiquette [involved in excusing himself from the table]."

As a result of his over concern for etiquette, Tycho became ill and unable to urinate. The problem persisted for five days. He began to feel better, except for a continued fever and an inability to sleep. Soon, he became delirious and refused to eat. On the night of October 24, he was heard yelling out repeatedly that he hoped his life had not been lived in vain. The next morning he was no longer delirious, but he felt weak and realized he was near death. His eldest son and second daughter and her husband were not in Prague. He asked his younger son and his pupils to keep studying the heavenly bodies and begged Kepler to finish the *Rudolphine Tables*.

Tycho Brahe died on October 24, 1601, at the age of fifty-four years and ten months. It is uncertain exactly what caused his death. Most modern medical authorities believe he suffered from a swollen prostate.

In a lavish funeral procession on November 4, people walked and carried candles decorated with the arms of the Brahe family, followed by a banner of black with the arms and name of Tycho Brahe. Next came his favorite horse, another banner, and another horse. Other marchers brought up the rear carrying a helmet with the family colors, a pair of spurs, and a shield with the Brahe coat of arms. Last came the coffin, followed by friends and family.

Part of the eulogy dealt with how Tycho had always said what he felt, which sometimes earned him enemies. Mostly, though, the speech highlighted his scientific achievements. A monument was eventually erected over his tomb in the crypt of Teyn Church in Prague that consisted of a full figure in relief of Tycho clad in armor. His left hand is on his sword hilt and his right on a globe. Underneath the globe is a shield with the arms of Brahe, Bille, and the families of his two grandmothers. A helmet

lies at his feet. The epitaph reads, "To be rather than to be perceived."

After Tycho's death, Emperor Rudolph granted Tycho's common-law wife and children noble status. The youngest daughter, Cecily, became a baroness. Magdalene never married after the discouraging experience with Gellius, but Elisabeth married the nobleman Frans Tengnagel, who had remained loyal to Tycho for more than six years. Before he died, Tycho became a grandfather, as Elisabeth had become pregnant before her marriage.

Tengnagel sold Tycho's instruments to Rudolph but never received payment; however, interest on the debt provided a good income for the family. The son of Tengnagel and Elisabeth, Rudolph Tengnagel, ended up in control of Hradcany Castle and all of Prague west of the Vltava River.

Tengnagel gave Kepler access to Tycho's observation records and worked with him to edit Tycho's unpublished work on the new star of 1572. The astronomer's scientific correspondence, and a reissue of his instrument book, were published between 1601 and 1602. At last, Tycho's works were before the public and available throughout Europe.

The publication of the instrument book revealed to a wider audience the advancements Tycho had made in designing and building astronomical instruments. He was the first to realize that a mural quadrant was more stable than an armillary. He preferred the mural quad-

This 1660 painting depicts the Tychonic System, which was adopted by many during the seventeenth century who were resistant to the idea of a moving Earth. In this illustration, Tycho is seated in the bottom right corner. *(Courtesy of The British Library.)*

This monument to Tycho Brahe was built after his death in 1601. *(Courtesy of the Library of Congress.)*

rant over the triquetrum for measuring altitudes because quadrants could also determine time. Tycho gave up depending on clocks because they were so unreliable. The main advantage of the sextant, invented in the Middle East but improved by Tycho, was that it could be disassembled and reassembled for easy transport.

His advancements in the use of instruments, and his adherence to strict protocol while performing experiments, meant that his astronomical tables were the most accurate ever. For example, his measurement of the tropical year—356 degrees, 5 hours, 48 minutes, 45 seconds—was off by only one second. His tables for the apparent motion of the Sun were the most accurate to date. He provided a new catalog of stars—the first since Ptolemy's, which had contained numerous calculation and copying errors.

Kepler had to fight Tycho's heirs almost continuously over the next decade in order to maintain control of Tycho's data. He was successful in doing so, often at great sacrifice, and for the first time, theoretical genius was married to astronomical measurements that could be trusted. As a result, his famous Three Laws of Planetary Motion marked the beginning of physical astronomy and started the process that culminated in Isaac Newton's discovery of gravity and his breakthrough in terrestrial and celestial physics.

When Tycho died, more than fifty former *familia* members continued working in Europe, however, none on the same scale as Tycho had. Johannes Kepler, for instance, employed no more than two assistants and made about one-sixth Tycho's salary. The Jesuits, a group within the Catholic Church that became prominent as educators over the next century, adopted and recast Tycho's methods to create a replica of Uraniborg in Beijing, China, in 1669. In Europe, the Leidin Observatory opened in 1632, as did the Royal Stjerneborg Observatory at the University of Copenhagen in 1637 (with Longomontanus's help). However, most of these observatories, including the ones in London and Paris,

A Tychonic equatorial armillary was erected in Beijing, China, in 1669, by members of the Society of Jesus, or the Jesuits. *(Courtesy Cambridge University Press.)*

carried out small-scale, individual research efforts.

In 1901, the remains of Tycho and his wife were exhumed. His were mostly intact, except that his gold and silver nose piece had disappeared, probably stolen by gravediggers. As one witness reported:

> The body in the coffin resembled the . . . monument [outside the tomb] to an extraordinary degree. The head was slightly turned to one side, the bones of the face and the peaked Spanish beard were well preserved. The head was covered by a skull-cap, and the neck

surrounded by a Spanish ruff, which, like the remainder of the clothing, had suffered little during the 300 years.

In the late 1900s, Uraniborg was partly rebuilt. A quarter of the garden was replanted in the way it had appeared in Tycho's day. Tourists can still walk through the ruins of Stjerneborg. While Hven remains a picturesque island in the summer months, a visitor will find only hints of the glory that prevailed there when Tycho ruled the land and studied the sky.

TIMELINE

1543   Nicholas Copernicus's book, *On the Revolution of the Celestial Spheres,* is published; Copernicus dies.

1546   Tycho Brahe born on December 14 in Helsingborg, Denmark; Martin Luther, founder of Protestant Reformation, dies.

1549   Kidnapped by his Uncle Jorgen and Aunt Inger.

1559   Enters the University of Copenhagen; Frederick II becomes king of Denmark.

1560   Tycho begins studying astronomy.

1562   Attends Leipzig University in Germany.

1563   Observes conjunction of Saturn and Jupiter; decides to become an astronomer.

1564   Birth of Galileo Galilei; birth of William Shakespeare; death of Michelangelo.

1565   Tycho returns to Denmark.

1566   Attends the University of Wittenberg in Germany; later attends the university in Rostock, Germany; witnesses lunar eclipse; loses nose in a duel; French astrologer Nostradamus dies.

1569 Tycho begins construction of Augsburg quadrant.

1570 Inherits half of Knudstrup castle after father's death.

1571 Birth of Johannes Kepler.

1572 Tycho discovers a new star.

1573 Publishes *The New Star* (*De stella nova*); Francis Drake sees Pacific Ocean for first time.

1575 Tycho meets with Frederick II, king of Denmark; receives ownership of Hven.

1576 Begins construction of Uraniborg.

1577 Discovers a comet; begins work on *Stella Caudata* (*The Star with a Tail*); begins development of Tychonic system; Christian IV of Denmark is born.

1582 Tycho builds mural quadrant.

1584 Builds Stjerneborg.

1588 Frederick II dies.

1589 Galileo becomes professor of mathematics at the University of Pisa.

1596 Christian IV becomes king of Denmark; Johannes Kepler publishes *Cosmological Mystery*.

1597 Tycho leaves Uraniborg; moves to Germany.

1598 Finishes his book concerning his new instruments, *Astronomiae instauratae mechanica*.

1599 Meets Rudolph II, emperor of the Holy Roman Empire; receives appointment as imperial mathematician; moves to Benátky castle.

1600 Johannas Kepler begins work for Tycho; Tycho and family move to Prague.

1601 Tycho Brahe dies on October 24, 1601.

SOURCES

CHAPTER ONE: Noble Genius

p. 12, "something divine that men . . ." J.L.E. Dreyer, *Tycho Brahe: A Picture of Scientific Life and Work in the Sixteenth Century* (1890; reprint, New York: Dover, 1963), 14.

CHAPTER TWO: Student Days

p. 32, "Neither my country nor . . ." John Allyne Gade, *The Life and Times of Tycho Brahe* (Princeton, NJ: Princeton University Press, 1947), 30.

p. 33, "[Brahe] unexpectedly got into . . ." Dreyer, *A Picture of Life and Work*, 44.

p. 34, "As Tycho was not used . . ." Ibid.

p. 36, "I was better received..." Victor E. Thoren, *The Lord of Uraniborg: A Biography of Tycho Brahe* (Cambridge, UK: Cambridge University Press, 1990), 28.

p. 39, "an astronomer, more than the student . . ." John Robert Christianson, *On Tycho's Island: Tycho Brahe and His Assistants, 1570-1601* (Cambridge, UK: Cambridge UP, 2000), 8.

CHAPTER THREE: Stargazer Receives an Offer

p. 42, "The woman who for three winters . . ." Thoren, *The Lord of Uraniborg*, 46.

p. 46, "of such value that . . ." John Allyne Gade, *The Life and Times of Tycho Brahe* (Princeton, NJ: Princeton University Press, 1947), 53-54.

p. 46, "horses, dogs, and luxury" Arthur Koestler, *The Sleepwalkers: A History of Man's Changing Vision of the Universe* (New York: Macmillan, 1959), 288.

p. 47, "feats of arms, concourse . . ." Thoren, *The Lord of Uraniborg*, 71.

CHAPTER FOUR: His Lordship of Uraniborg

p. 49, "One of my . . ." Christianson, *On Tycho's Island*, 22-23.

p. 50, "I saw the little island . . ." Ibid., 12.

p. 50, "Apollo desires it . . ." Thoren, *The Lord of Uraniborg*, 104.

p. 51, "with its white cliffs . . ." Dreyer, *A Picture of Life and Work*, 88.

p. 53, "We, Frederick the Second . . ." Ibid., 86-87.

CHAPTER FIVE: Life on Hven

p. 65, "See how your people are laving . . ." Dreyer, *A Picture of Life and Work*, 128.

p. 65, "The squire is on land!" Ibid.

CHAPTER SIX: The Star with a Tail

p. 75, "childish and doubtful" Koestler, *The Sleepwalkers*, 294.

p. 75, "juvenile and habitually mediocre" Ibid.

p. 75, "virile, precise . . ." Ibid.

CHAPTER SEVEN: Calamity

p. 86, "Muses' royal castle..." Christianson, *On Tycho's Island*, 141.

p. 89, "if his grace could have ..." Dreyer, *A Picture of Life and Work*, 156-157.

p. 90, "unpleasant obstacles" Ibid., 217.

p. 94, "How happy I am ..." Gade, *Life and Times of Tycho Brahe*, 142.

CHAPTER EIGHT: Starting Over

p. 98, "was dismissed in disgrace ..." Dreyer, *A Picture of Life and Work*, 236.

p. 99, "exile over dishonor" Ibid., 204.

p. 100, "But it has turned out differently ..." Ibid., 244.

p. 100, "if it could be done ..." Ibid., 244.

p. 100, "It is by no means from any ..." Ibid., 245.

p. 102, "Your letter is somewhat peculiarly ..." Thoren, *The Lord of Uraniborg*, 380.

CHAPTER NINE: Enter Kepler

p. 105, "have learned that this is ..." Edward Rosen, *Three Imperial Mathematicians: Kepler Trapped Between Tycho Brahe and Ursus* (New York: Abaris, 1986), 35.

p. 105, "has not mastered the precise ..." Ibid., 35.

p. 106, "his greatness" Ibid., 106.

p. 106, "feeble, inept, and childish" Ibid., 106.

p. 106, "I do not know by what chance ..." Ibid., 110.

p. 108, "snatched and stolen from others ..." Ibid., 112.

p. 108, "done so to excess!" Ibid., 112.

p. 108, But you wrote these things ..." Ibid., 113.

CHAPTER TEN: Rebirth

p. 112, "I went in . . ." Christianson, *On Tycho's Island*, 234-236.

p. 115, "Tycho possesses . . ." Max Caspar, *Kepler*, trans. and ed. C. Doris Hellman (New York: Dover, 1993), 102-103.

p. 118, "the feebleness of old age" Ibid. 307.

CHAPTER ELEVEN: End of an Era

p. 119, "Holding his urine . . ." Rosen, *Three Imperial Mathematicians*, 313.

p. 120, "To be rather than . . ." Christianson, *On Tycho's Island*, 92.

p. 124, "The body in the coffin . . ." Gade, *The Life and Times*, 188.

BIBLIOGRAPHY

Caspar, Max. *Kepler*. Translated and edited by C. Doris Hellman. New York: Dover, 1993.

Christianson, John Robert. *On Tycho's Island: Tycho Brahe and His Assistants, 1570-1601*. Cambridge, UK: Cambridge UP, 2000.

Dreyer, J.L.E. *Tycho Brahe: A Picture of Scientific Life and Work in the Sixteenth Century*. 1890. Reprint, New York: Dover, 1963.

————. *A History of Astronomy from Thales to Kepler*. New York: Dover, 1953.

Gade, John Allyne. *The Life and Times of Tycho Brahe*. Princeton, NJ: Princeton University Press, 1947.

Gingerich, Owen. *The Great Copernicus Chase and Other Adventures in Astronomical History*. Cambridge, Mass., Sky Publishing, 1992.

————. *The Eye of Heaven: Ptolemy, Copernicus, Kepler*. New York: American Institute of Physics, 1993.

Koestler, Arthur. *The Sleepwalkers: A History of Man's Changing Vision of the Universe*. New York: Macmillan, 1959.

Rosen, Edward. *Three Imperial Mathematicians: Kepler Trapped Between Tycho Brahe and Ursus*. New York: Abaris, 1986.

Thoren, Victor E. *The Lord of Uraniborg: A Biography of Tycho Brahe*. Cambridge, UK: Cambridge UP, 1990.

WEBSITES

Landskrona Cultural Department, Sweden: Tycho Brahe Website

http://www.tychobrahe.com/

Museum of the History of Science, Oxford: Images of Tycho Brahe

http://www.mhs.ox.ac.uk/tycho/index.htm

Rice University: The Galileo Project

http://es.rice.edu/ES/humsoc/Galileo/People/tycho_brahe.html

The Royal Library, Copenhagen: The Instruments of Tycho Brahe

http://www.kb.dk/elib/lit/dan/brahe/index-en.htm

INDEX